Linux for Makers

Understanding the Operating System That Runs Raspberry Pi and Other Maker SBCs

Aaron Newcomb

MAKER**MEDIA**

SAN FRANCISCO, CA

Linux for Makers

by Aaron Newcomb

Published by Maker Media, Inc., 1160 Battery Street East, Suite 125, San Francisco, CA 94111.

Maker Media books may be purchased for educational, business, or sales promotional use. Online editions are also available for most titles (*http://oreilly.com/safari*). For more information, contact our corporate/institutional sales department: 800-998-9938 or *corporate@oreilly.com*.

Editor: Patrick DiJusto
Production Editor: Kristen Brown
Copyeditor: Gillian McGarvey
Proofreader: Rachel Monaghan
Indexer: Angela Howard
Interior Designer: David Futato
Cover Designer: Brian Jepson
Illustrator: Rebecca Demarest

May 2017:   First Edition

Revision History for the First Edition

2017-04-05:   First Release

See *http://oreilly.com/catalog/errata.csp?isbn=9781680451832* for release details.

978-1-680-45183-2

[LSI]

Contents

Preface

When I started a Makerspace in my local community, I noticed some particularly interesting learning trends. Some people were reluctant to learn a new skill until someone shared some basic techniques that helped get them started down the path of understanding. Other users would jump right into learning a new skill without any idea of what they were doing. This would lead to slow progress until, again, someone provided some assistance that would lead them in the right direction. In both cases, just a little guidance in the beginning greatly accelerated the learning process.

Learning how to use Linux for making and building projects is no easy task. In many cases, Makers cut and paste from a website tutorial into the Linux command line without understanding what they are actually doing, only to be frustrated when they want to modify or tweak something to suit their needs. Also, many Makers shy away from using the Raspberry Pi or similar boards because they feel Linux is too foreign and that using a command line as indicated in many tutorials will be more difficult than using a GUI.

This book aims to overcome those fears and provide a foundation for further learning and exploration when using the Linux operating system for your projects. Linux is just another tool in your Maker tool belt. It might be different from other operating systems you've used in the past, but—like all tools—it's no more challenging to use once you know how to use it effectively. In fact, Linux is so powerful, you may start to prefer it to other operating systems and choose to use it on a daily basis.

Linux is a powerful open source operating system that has been around for many years and is widely used for running servers and websites. But most students and Makers encounter it for the first time when they're working on projects with their Raspberry Pi or similar single-board computer (SBC), such as

BeagleBone Black or Intel Galileo. *Linux for Makers* is the first book that explains the Linux operating system specifically for Makers as opposed to programmers and administrators. By gaining a deeper understanding of Linux, Makers can add another useful tool to their kit that will help them build projects more easily.

Because this book was written with today's Maker in mind, it will focus mostly on the Raspbian distribution of Linux running on the Raspberry Pi, as that platform is the most prolific in the ecosystem today. However, most of the topics covered will apply broadly to other Linux distributions, and I will indicate when they may differ. To that end, this book will focus on the basic principles that a Maker needs to know, avoiding details that are not particularly relevant to building projects. After loading the operating system, I will cover the principles of how Linux works, how to use the command line, how to control devices, and loads of tips and tricks that can help you be more effective.

Throughout the book, you will find sections called "Try It for Yourself" where you can get your hands dirty practicing what you just learned and explore additional opportunities to try out new concepts. I have also included illustrations and pictures throughout the book that should help clarify what you can expect to see as you use Linux on your Raspberry Pi.

I have also included a brief history of Linux in Appendix A for those readers who might be wondering "How did all this get started?" or "How did Linux end up getting put together the way it did?"

Conventions Used in This Book

The following typographical conventions are used in this book:

Italic
> Indicates new terms, URLs, email addresses, filenames, and file extensions.

`Constant width`
> Used for program listings, as well as within paragraphs to refer to program elements such as variable or function

names, databases, data types, environment variables, statements, and keywords.

Constant width bold
Shows commands or other text that should be typed literally by the user.

Constant width italic
Shows text that should be replaced with user-supplied values or by values determined by context.

 This element signifies a tip, suggestion, or a general note.

 This element indicates a warning or caution.

O'Reilly Safari

 Safari (formerly Safari Books Online) is a membership-based training and reference platform for enterprise, government, educators, and individuals.

Members have access to thousands of books, training videos, Learning Paths, interactive tutorials, and curated playlists from over 250 publishers, including O'Reilly Media, Harvard Business Review, Prentice Hall Professional, Addison-Wesley Professional, Microsoft Press, Sams, Que, Peachpit Press, Adobe, Focal Press, Cisco Press, John Wiley & Sons, Syngress, Morgan Kaufmann, IBM Redbooks, Packt, Adobe Press, FT Press, Apress, Manning, New Riders, McGraw-Hill, Jones & Bartlett, and Course Technology, among others.

For more information, please visit *http://oreilly.com/safari*.

How to Contact Us

Please address comments and questions concerning this book to the publisher:

Make:
1160 Battery Street East, Suite 125
San Francisco, CA 94111
877-306-6253 (in the United States or Canada)
707-639-1355 (international or local)

We have a web page for this book, where we list errata, examples, and additional information. You can access this page at *http://bit.ly/linux_for_makers*.

Make: unites, inspires, informs, and entertains a growing community of resourceful people who undertake amazing projects in their backyards, basements, and garages. Make: celebrates your right to tweak, hack, and bend any technology to your will. The Make: audience continues to be a growing culture and community that believes in bettering ourselves, our environment, our educational system—our entire world. This is much more than an audience; it's a worldwide movement that Make is leading. We call it the Maker Movement.

For more information about Make:, visit us online:

Make: magazine: *http://makezine.com/magazine*
Maker Faire: *http://makerfaire.com*
Makezine.com: *http://makezine.com*
Maker Shed: *http://makershed.com*

To comment or ask technical questions about this book, send email to *bookquestions@oreilly.com*.

Acknowledgments

I would like to thank my wife Jennifer and kids Stephen, Olivia, and James for being so patient with me as I wrote this book. Many nights and weekends were taken out of my already busy schedule to work on it, and they were supportive through it all.

Thanks to James who introduced my to Linux back in 1997. Know someone who might like Linux or Raspberry Pi? Tell them about it!

I am grateful to the support of my editor, Patrick, and all the staff at Maker Media and O'Reilly Media who guided me through the writing, editing, and reviewing process.

I also want to add a big shoutout to the people who gave up their time to help review the book and offer so many great suggestions—Robert Shaver, Christoph Zimmermann, Jim Kennon, Rashed Harun, and Broedy Bowers.

1/Getting Started

The Raspberry Pi is a single-board computer (SBC), which means that—as the name suggests—it is a complete computer system built on a single printed circuit board (PCB). Like most SBCs, it doesn't come out of the box ready to power up and use. It has the same basic components built into the board as any other computer has: a central processing unit (CPU), memory, video processor, audio, and networking.

The one thing it doesn't have out of the box is a storage device. On a computer or laptop, most people use a hard drive that contains the operating system and all their files. With a Raspberry Pi, you use an SD card as the main storage device. So before we can tackle the ins and outs of making things using Linux, we need to load the operating system you want to use with the Raspberry Pi on the SD card. For best results, you should use an SD card that has at least 8 GB of available capacity.

What Is a Disk Image?

A *disk image* is a single file that represents a point-in-time copy of an entire storage device. Just like a photograph is an image that can contain different people or objects, a disk image can contain lots of different partitions, directories, and files.

This process can be confusing for some people, so let's break it down into individual steps and take them one at a time. These steps include downloading a compressed disk image from the internet, uncompressing that image on your local computer, writing that image to your SD card, and finally, booting up your Raspberry Pi. You will find that these steps apply to other SBCs as well, although the exact image file you use will change because the operating system needs to be built specifically for the board you're using.

I will be mentioning some concepts in this chapter that may be new to you, like filesystems, terminal emulators, and the command line. Don't worry. We will cover these in depth in Chapter 2. For now, we just need to get things running so you can use your Raspberry Pi as you go through this book. In order to get started, you will need access to a desktop or laptop computer and a connection to the internet.

Choosing and Downloading a Disk Image

The best place to find the most up-to-date disk images for the Raspberry Pi is on the Raspberry Pi Foundation website at *http://raspberrypi.org/downloads*. When you get to that site, you will see that there are a number of disk images to choose from. The two that have official support from the Raspberry Pi Foundation are NOOBS and Raspbian. *NOOBS* is essentially an installer for Raspbian as well as several other operating systems that can be run on the Raspberry Pi. NOOBS automates several of the tasks described in this chapter, but is not the actual operating system you will end up using. Instead, the operating system is downloaded as part of the installation process. *Raspbian*, on the other hand, contains the actual operating system, and the installation method requires you to manually write the disk image to an SD card.

I recommend choosing the Raspbian image for a number of reasons. First, while installing NOOBS might seem less complicated, once you learn the standard installation procedure, Raspbian will actually be easier and faster than using NOOBS. Second, learning the installation process for Raspbian will teach you how to load any disk image you might want to try in the future, and learning how to do things is what this book is all about. Third, since loading Raspbian using this process results in a more standard disk layout and structure on the SD card, it makes it a little easier to back up your image for safekeeping. This is something you definitely want to do to protect your work and prevent data loss in case you break your SD card or damage your Raspberry Pi.

Click on the Raspian link and download the Raspbian Jessie file. Later, if you decide you don't want to use the desktop (see Chapter 3) you can choose the Raspbian Jessie Lite file, which is smaller in size but does not come prepackaged with a desktop environment. Save the ZIP file in a place you can find easily later, like the *Downloads* folder. At the time of this writing the filename looks like this: *2017-01-11-raspbian-jessie.zip*.

Uncompressing the Disk Image

Disk images are compressed to make them smaller for downloading. Compressing a file basically uses an algorithm to take out all the empty spaces and duplicate existing information. For example, imagine removing all the spaces between the words of an essay or a text to a friend. It makes the finished product much smaller, but also much more difficult to read. So in order to make use of the disk image, you will first need to uncompress it. You will need software in order to uncompress the disk image, and in most cases that software is already available on your desktop or laptop.

Windows

In recent versions of Windows, the decompression software you need is already built into File Explorer. Simply open it up and navigate to the folder where you saved the ZIP file. Double-click on the ZIP file to open it. Clicking on the Extract button will save the disk image to a place of your choosing that you can easily find later. This could be the same folder where you saved the ZIP file originally (see Figure 1-1 for an example of what this looks like).

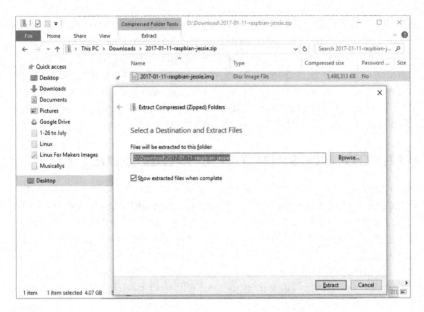

Figure 1-1. *Extracting the disk image using Windows*

MacOS

When you download the ZIP file with a recent version of Safari, the ZIP file will be automatically uncompressed and saved in your *Downloads* folder by default. If you are using an older version, macOS has a built-in tool for uncompressing files called Archive Utility. Just double-click on the ZIP file you downloaded, and Archive Utility will uncompress the file for you (see Figure 1-2).

When the process is complete. you will see the extracted *.img* file appear on your desktop or wherever you saved your ZIP file.

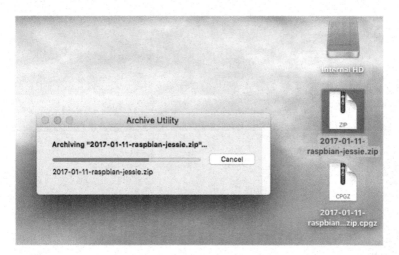

Figure 1-2. *The Archive Utility running on macOS*

Linux

Most distributions of Linux also come with built-in programs to extract compressed files. On the desktop, you can open the file browser and double-click on the ZIP file you downloaded. This will open up Archive Manager and allow you to extract the image file (see Figure 1-3).

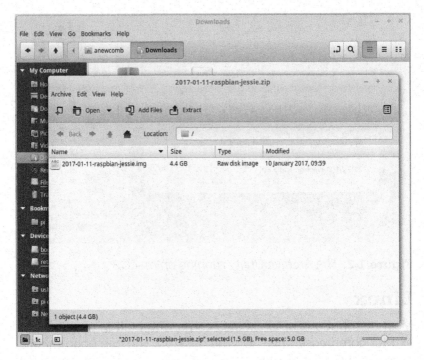

Figure 1-3. *Extracting the disk image using Linux*

You can also do this just as easily from the command line. Open a terminal emulator and type:

```
cd Downloads
unzip 2017-01-11-raspbian-jessie.zip
```

This assumes that you saved the ZIP file to your default download location and that the filename is correct. Be patient. This is a big file and uncompressing it will take some time. See Figure 1-4 to see what this looks like.

Command-Line Confusion

Not comfortable using the command line yet? That's a big part of what this book is about. You will learn more about that in the next few chapters.

```
anewcomb@anewcomb-VirtualBox ~ $ cd Downloads
anewcomb@anewcomb-VirtualBox ~/Downloads $ unzip 2017-01-11-raspbian-jessie.zip
Archive:  2017-01-11-raspbian-jessie.zip
  inflating: 2017-01-11-raspbian-jessie.img
anewcomb@anewcomb-VirtualBox ~/Downloads $ █
```

Figure 1-4. *Extracting the disk image using the Linux command line*

Writing the Disk Image to the SD Card

Be aware that after this step, any files you may have had on your SD card will be deleted.

Windows

Windows is currently the easiest operating system to use when writing a disk image to a card. However, unlike macOS and Linux, it requires you to download some software first. Open your browser and download the Win32 Disk Imager application (*https://sourceforge.net/projects/win32diskimager/*).

Install the application by double-clicking on the file you downloaded. After it installs successfully, connect your SD card to your computer and make note of the drive letter that Windows assigns to it. Now, open the application. The first thing to check is that the drive letter the application selects is actually the drive letter that corresponds to your SD card. Win32 Disk Imager is pretty good about only selecting SD cards, but always double-check this because you are about to erase the drive, and you definitely don't want to erase your *C:* drive on Windows.

Now click on the blue folder icon to select the image file you extracted in the previous step. See Figure 1-5 for an example of what this should look like.

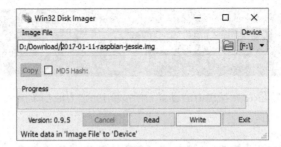

Figure 1-5. *The Win32 Disk Imager program interface*

When you've selected the correct file, you can click the Write button. This will overwrite all the data on the SD card. When the process is completed, close the Win32 Disk Imager application. Open File Explorer, right-click on the SD card drive letter, and select Eject. Always do this before removing your SD card to make sure the computer is done writing files in the background.

MacOS

Like Linux, macOS already has all the software you need to write the image to an SD card. However, you will need to use the command line. Open Finder and select Applications→Utilities→Terminal (see Figure 1-6).

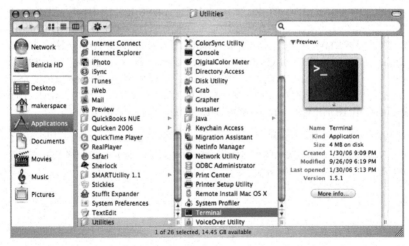

Figure 1-6. *Finding the Terminal program on macOS*

Insert the SD card and wait a few moments for macOS to recognize it. Use the `diskutil list` command in the terminal window to print a list of all the disks attached to your Mac:

```
diskutil list
```

Identify the disk (not partition) representing your SD card (e.g., disk1, not disk1s1) as shown in Figure 1-7.

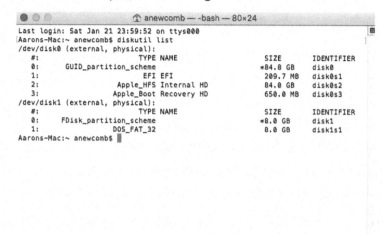

Figure 1-7. *Example output of the diskutil list command*

In this case, I have a 64 GB SD card and macOS recognizes it as disk1.

Unmount your SD card by using the `diskutil unmountDisk` command and the disk name to prepare for copying data to it (see Figure 1-8:

```
diskutil unmountDisk /dev/disk1
```

```
●  ●  ●                    ⚙ anewcomb — -bash — 80×24
Last login: Sat Jan 21 23:59:52 on ttys000
Aarons-Mac:~ anewcomb$ diskutil list
/dev/disk0 (external, physical):
   #:                       TYPE NAME                    SIZE       IDENTIFIER
   0:      GUID_partition_scheme                        *84.8 GB    disk0
   1:                        EFI EFI                     209.7 MB    disk0s1
   2:                  Apple_HFS Internal HD             84.0 GB     disk0s2
   3:                 Apple_Boot Recovery HD             650.0 MB    disk0s3
/dev/disk1 (external, physical):
   #:                       TYPE NAME                    SIZE       IDENTIFIER
   0:     FDisk_partition_scheme                        *8.0 GB     disk1
   1:                 DOS_FAT_32                          8.0 GB     disk1s1
Aarons-Mac:~ anewcomb$ diskutil unmountDisk /dev/disk1
Unmount of all volumes on disk1 was successful
Aarons-Mac:~ anewcomb$ ▊
```

Figure 1-8. *Using diskutil to unmount a disk*

Now we need to copy the image file over to the SD card. We will use the *data duplicator*, or **dd** command, for this. Be very careful to get the right disk number so that you don't overwrite your system disk! You will also need to use *SuperUserDO* (**sudo**) for this command. **sudo** is a secure way for a regular user to run a command that normally requires administrator privileges:

```
sudo dd if=Desktop/2017-01-11-raspbian-jessie.img of=/dev/
rdisk1 bs=1m
```

You will be asked for your password, since you are using **sudo**. This may result in an error if you have GNU coreutils installed:

```
dd: invalid number '1m'
```

Don't worry about what that means just yet—all you need to do is use a block size of 1m in the **bs=** section of the command, as follows:

```
sudo dd if=Desktop/2017-01-11-raspbian-jessie.img of=/dev/
rdisk1 bs=1m
```

This will take quite a few minutes, depending on the image file size. You can check the progress by pressing Ctrl-T to send a SIGINFO signal (see Figure 1-9). If this command still fails, try using the **disk** command instead of **rdisk**.

```
● ● ●                    🐧 anewcomb — dd ‹ sudo — 80×29
Last login: Sat Jan 21 23:59:52 on ttys000
Aarons-Mac:~ anewcomb$ diskutil list
/dev/disk0 (external, physical):
   #:                       TYPE NAME                    SIZE       IDENTIFIER
   0:      GUID_partition_scheme                        *84.8 GB    disk0
   1:                        EFI EFI                     209.7 MB    disk0s1
   2:          Apple_HFS Internal HD                     84.0 GB     disk0s2
   3:          Apple_Boot Recovery HD                    650.0 MB    disk0s3
/dev/disk1 (external, physical):
   #:                       TYPE NAME                    SIZE       IDENTIFIER
   0:     FDisk_partition_scheme                        *8.0 GB     disk1
   1:                 DOS_FAT_32                          8.0 GB     disk1s1
Aarons-Mac:~ anewcomb$ diskutil unmountDisk /dev/disk1
Unmount of all volumes on disk1 was successful
Aarons-Mac:~ anewcomb$ sudo dd if=Downloads/2017-01-11-raspbian-jessie-lite.img
of=/dev/rdisk1 bs=1m

WARNING: Improper use of the sudo command could lead to data loss
or the deletion of important system files. Please double-check your
typing when using sudo. Type "man sudo" for more information.

To proceed, enter your password, or type Ctrl-C to abort.

Password:
load: 1.37  cmd: dd 538 uninterruptible 0.00u 0.60s
549+0 records in
548+0 records out
574619648 bytes transferred in 71.079200 secs (8084217 bytes/sec)
█
```

Figure 1-9. *Using the dd utility on macOS*

In this case, the dd command has transferred 549 1 MB blocks. Once the process is complete, the terminal will bring you back to a prompt. Finally, run one last command before you disconnect your SD card:

    sync

This will make sure all the writes to the SD card that may be occurring in the background have finished. You can now remove your SD card.

Linux

Before you connect your SD card to your computer, run this command:

    sudo fdisk -l

This will show you all the physical drives connected to your system. Note the size of the drive and the name of the disk. See Figure 1-10.

```
anewcomb@anewcomb-VirtualBox ~ $ sudo fdisk -l

Disk /dev/sda: 21.5 GB  21474836480 bytes
255 heads, 63 sectors/track, 2610 cylinders, total 41943040 sectors
Units = sectors of 1 * 512 = 512 bytes
Sector size (logical/physical): 512 bytes / 512 bytes
I/O size (minimum/optimal): 512 bytes / 512 bytes
Disk identifier: 0x000c6e58

   Device Boot      Start         End      Blocks   Id  System
/dev/sda1   *        2048    37748735    18873344   83  Linux
/dev/sda2        37750782    41940991     2095105    5  Extended
/dev/sda5        37750784    41940991     2095104   82  Linux swap / Solaris
anewcomb@anewcomb-VirtualBox ~ $ ▌
```

Figure 1-10. *Using fdisk to find your physical disks*

Now connect your SD card to your computer and wait a few moments for your Linux PC to recognize it. It may automatically mount and display any existing partitions on the drive. Now run the same command again. This time you should notice a new drive. (See Figure 1-11.) Note the disk name and size. The disk name will be important for the next step.

```
anewcomb@anewcomb-VirtualBox ~ $ sudo fdisk -l

Disk /dev/sda: 21.5 GB, 21474836480 bytes
255 heads, 63 sectors/track, 2610 cylinders, total 41943040 sectors
Units = sectors of 1 * 512 = 512 bytes
Sector size (logical/physical): 512 bytes / 512 bytes
I/O size (minimum/optimal): 512 bytes / 512 bytes
Disk identifier: 0x000c6e58

   Device Boot      Start         End      Blocks   Id  System
/dev/sda1   *        2048    37748735    18873344   83  Linux
/dev/sda2        37750782    41940991     2095105    5  Extended
/dev/sda5        37750784    41940991     2095104   82  Linux swap / Solaris

Disk /dev/sdb: 7985 MB, 7985954816 bytes
231 heads, 28 sectors/track, 2411 cylinders, total 15597568 sectors
Units = sectors of 1 * 512 = 512 bytes
Sector size (logical/physical): 512 bytes / 512 bytes
I/O size (minimum/optimal): 512 bytes / 512 bytes
Disk identifier: 0x000cc086

   Device Boot      Start         End      Blocks   Id  System
/dev/sdb1            2048    15597567     7797760    b  W95 FAT32
anewcomb@anewcomb-VirtualBox ~ $ ▌
```

Figure 1-11. *Using fdisk to find your physical disks (continued)*

Now run the following commands to write the image file to your SD card. Be very careful to use the SD card disk name and not your system disk:

```
sudo umount /dev/YourSDCardName*
```

Replace */YourSDCardName* with the disk name you identified as your SD card with a * on the end:

```
cd ~/Downloads
```

```
sudo dd if=2017-01-11-raspbian-jessie.img of=/dev/sdb bs=4M
```

Of course, substitute your filename in the `if` = part of the `dd` command and substitute the disk name of your SD card in the `of=` part. This last command could take up to 10 minutes to complete depending on the speed of your computer and SD card. Once it is complete, the prompt will return (see Figure 1-12).

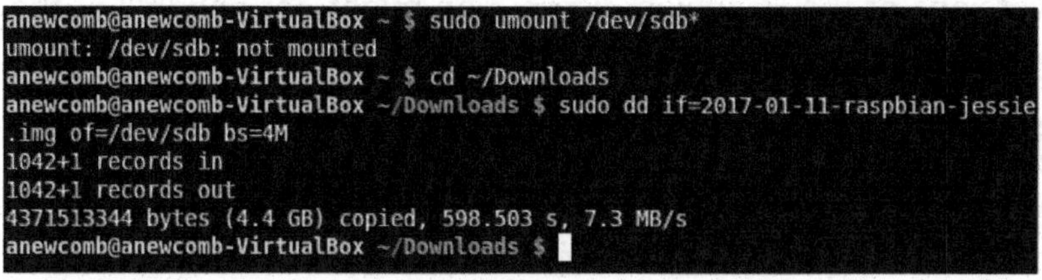

```
anewcomb@anewcomb-VirtualBox ~ $ sudo umount /dev/sdb*
umount: /dev/sdb: not mounted
anewcomb@anewcomb-VirtualBox ~ $ cd ~/Downloads
anewcomb@anewcomb-VirtualBox ~/Downloads $ sudo dd if=2017-01-11-raspbian-jessie
.img of=/dev/sdb bs=4M
1042+1 records in
1042+1 records out
4371513344 bytes (4.4 GB) copied, 598.503 s, 7.3 MB/s
anewcomb@anewcomb-VirtualBox ~/Downloads $
```

Figure 1-12. *Using the dd utility on Linux*

Finally, run one last command before you disconnect your SD card:

```
sync
```

This will make sure all the writes to the SD card that may be occurring in the background have finished. You can now remove your SD card.

Booting the Raspberry Pi for the First Time

Now the magical time has come. You can insert your SD card into the Raspberry Pi, but before you connect the power, be sure your display is connected and that you're using a power supply that is at least 1 amp (usually displayed on the power supply as 1A) or larger. Two amps (2A) is actually better. Newer smartphone chargers should work, and connecting the Pi to a laptop or PC via USB should also be adequate.

As your Raspberry Pi boots up, you will notice four Raspberry Pi logos at the top of the screen and a bunch of scrolling text on a black screen with a lot of green OK text on each line. Don't be alarmed. This is normal. Linux is just starting up the operating system and launching some services (more about this in Chapter 2).

When your Raspberry Pi finishes booting up, there are a few settings you might want to adjust before you begin using it for the rest of this book. Other SBCs may or may not require these adjustments. Also, keep in mind that the developers of the operating system are updating the system all the time. Later versions of the Raspbian distribution might not need all of the suggestions that follow here.

Expanding the Filesystem

When that image file you downloaded was created, the people who made it were nice enough to shrink it down first to make the download faster. When you first boot up your system, you only have a total of 4 GB of space and only about 700 MB available to you even though your actual SD card is much bigger than that. Now that you have the image loaded onto your SD card, you probably want to expand the filesystem to get access to all that extra space. In the latest version of Raspbian, this happens automatically on your first boot. You will notice that on the first boot, your system will reboot after a short time. This is normal since the system has to reboot for it to recognize the new size of the filesystem. For older versions of Raspbian, on the Raspberry Pi desktop, click on the icon that looks like a dark monitor. This will open up a *terminal emulator,* which we'll use to access the command line. In the window that pops up, run this command:

```
sudo raspi-config
```

This will open up the Raspberry Pi configuration application as shown in Figure 1-13.

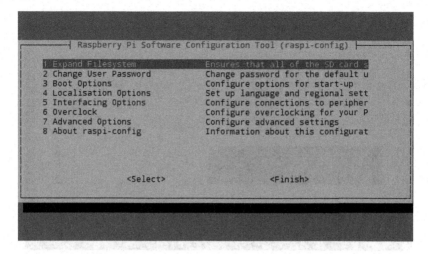

```
┌──────────┤ Raspberry Pi Software Configuration Tool (raspi-config) ├──────────┐
│                                                                                │
│    1 Expand Filesystem          Ensures that all of the SD card s              │
│    2 Change User Password       Change password for the default u             │
│    3 Boot Options               Configure options for start-up                │
│    4 Localisation Options       Set up language and regional sett             │
│    5 Interfacing Options        Configure connections to peripher             │
│    6 Overclock                  Configure overclocking for your P             │
│    7 Advanced Options           Configure advanced settings                   │
│    8 About raspi-config         Information about this configurat             │
│                                                                                │
│                                                                                │
│                                                                                │
│                  <Select>                        <Finish>                      │
│                                                                                │
└────────────────────────────────────────────────────────────────────────────┘
```

Figure 1-13. *The raspi-config screen*

Press the Enter key to select the Expand Filesystem option. After a few seconds, another screen will appear telling you that the root filesystem has been resized.

Changing the Localization Options

By default, the Raspbian image comes set up for use in the United Kingdom, since that is where the Raspberry Pi Foundation is located. If you live in another country, you will want to set up your Pi to use your local time zone, keyboard layout, and language settings. Believe me, it can be quite frustrating to type a double-quote character on your keyboard and have the Raspberry Pi interpret that as an @ symbol.

From the same program, use the down arrow on your keyboard to select Localisation Options or press the corresponding number key. We will be configuring all three options on this menu, but for now let's start with Change Locale. Press the Enter key on this option to continue. You should see a screen similar to Figure 1-14.

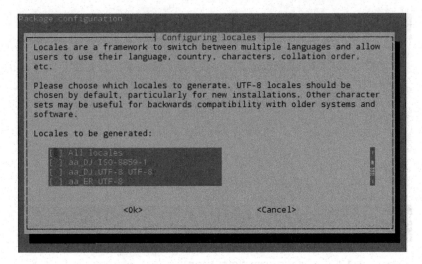

Figure 1-14. *The locales menu in raspi-config*

It might be tempting just to select "All locales" from this screen, but that would not be a good idea. Your Raspberry Pi would spend a very long time configuring the settings for hundreds of languages that you will probably never use. Instead, scroll through the choices and pick the locale that matches the language you speak most often. As the description suggests, choose a selection that ends in UTF-8. For example, in the United States you would choose en_US.UTF-8 by pressing the space bar when the selection indicator is next to your choice. I won't go over what all these codes mean in this book, but you can find a list of language codes online (*https://en.wikipe dia.org/wiki/List_of_ISO_639-1_codes*).

Once you've selected your locale, press the Enter key to move to the next screen, which will ask you what the default locale should be. Use the arrow keys to select the locale you just chose and press the Enter key again. Your Raspberry Pi will generate the needed information and return you to the main Raspberry Pi configuration screen.

Now it's time to set your time zone. Select Localisation Options again from the main screen and choose Change Timezone. This will present a screen similar to Figure 1-15.

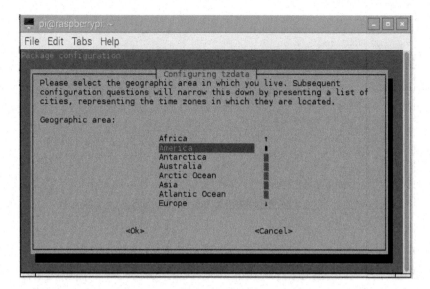

Figure 1-15. *Changing the time zone in raspi-config*

Use the arrow keys and Enter key on your keyboard to navigate this menu to select your time zone. If you're in the United States, you might find it easier to choose the America menu item. You can then type the first letter of the city closest to you to find the right time zone. Press the Enter key to confirm your selection and set your time zone. You will then be returned to the main Raspberry Pi configuration screen.

Last and probably most importantly, we need to set the keyboard layout. Select Localisation Options once again and select Change Keyboard Layout. You should see a screen similar to Figure 1-16.

Use the arrow keys to select the keyboard you are using. For most users, this will either be the Generic 104-key PC if you are in the US, or the Generic 105-key (Intl) PC if you are in Europe or elsewhere. Press the Enter key to move to the next screen.

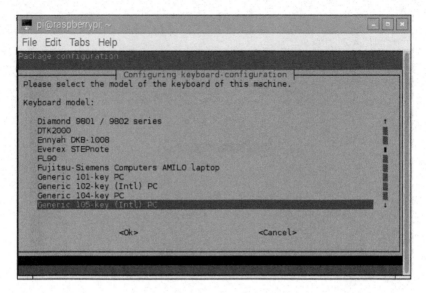

Figure 1-16. *Changing the keyboard layout in raspi-config*

The screen will list the English (UK) layouts that it was previously set up for. If you are not in the UK, you can find your keyboard layout by choosing Other from this menu and pressing the Enter key (see Figure 1-17).

On the next screen, choose the country of origin of your keyboard and press the Enter key. Now you can choose your layout. Use the arrow keys to move the cursor up to the top of the list, where you will find a generic layout you can choose, and press the Enter key to select it.

There are a few more screens to move through. Just select the defaults on those screens by pressing the Enter key for each one. You will then be returned to the main Raspberry Pi configuration screen.

Press the Tab key or use the arrow keys to access the Finish button and press the Enter key. You will be asked if you want to reboot now. Choose Yes and press Enter. Your Raspberry Pi will reboot.

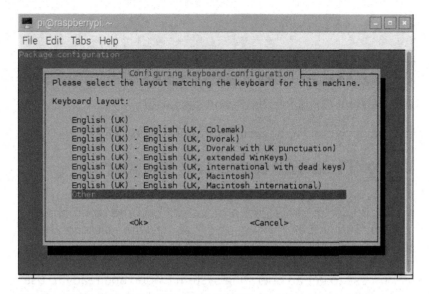

Figure 1-17. *Changing the keyboard layout in raspi-config (continued)*

Changing the Default Password

One last thing you should always do before you forget is to change the default password. You might think a small system like the Raspberry Pi couldn't be powerful enough to damage other systems on your network or the internet. This is incorrect. Even the smallest systems can be compromised to run and spread malicious software. This is especially true when you leave your password set to the default, as it will be the first thing someone will attempt to use when trying to break into your system.

Changing the password for the "pi" user is easily done via the command line. Open a terminal emulator window like you did previously and run the following command:

```
passwd
```

This will prompt you to enter the existing default password, which is "raspberry," and then enter a new password twice to make sure you have it correct (see Figure 1-18). Be sure to avoid easily guessed passwords like "password" or "12345678."

```
pi@raspberrypi:~ $ passwd
Changing password for pi.
(current) UNIX password:
Enter new UNIX password:
Retype new UNIX password:
passwd: password updated successfully
pi@raspberrypi:~ $ █
```

Figure 1-18. *Changing the default password*

You are now ready to begin using your Raspberry Pi!

Why This Matters for Makers

As you begin to use the Raspberry Pi and other SBCs more and more for your projects, this process will become commonplace for you. Remember, if things really go haywire, you can always start from scratch by following this process to load a new disk image onto your SD card. Many SBCs available today use a similar process to get an operating system loaded and ready to use, so learning these steps helps you prepare for future exploration of the many great boards out there.

2/Linux Principles

You are probably thinking "OK. Let's go! I am ready to begin my project." However, there are a few principles of the Linux operating system that you need to know about first. Using Linux in your project will be more complicated than using an Arduino. Linux is a full operating system with users, services, filesystems, and other resources that make it a very powerful and versatile platform for Makers. (By comparison, Arduino is based on a microcontroller with a limited set of instructions to execute.)

The Linux Desktop

A *graphical user interface* (GUI) is the way most people use their computers. Whether they use Windows, macOS, Android, or iOS (yes, mobile devices are just small computers), a GUI desktop is the canvas people use to make things happen. Linux is no exception here. Almost all Linux distributions come with a desktop environment to make interacting with programs easier and more functional. If you want to actively browse the web, create a document, or edit a photo, the desktop is the place to do it.

 Command Line Browser
You can check out the Lynx browser (*http:// lynx.browser.org/*) to see what it's like to browse the internet from the command line.

Single-board computers (SBCs) like the Raspberry Pi that run Linux are no exception. Figure 2-1 shows what the Raspberry Pi's desktop looks like at the time of this writing.

Figure 2-1. *The Raspbian desktop*

Just like Linux distributions (see Chapter 1), Linux desktops come in many flavors. The one used by default on Raspbian is called *Lightweight X11 Desktop Environment* (LXDE). *Lightweight* means it requires fewer resources to run and therefore works well on SBCs, which have less powerful CPUs and smaller memory footprints than modern desktops or laptops. Other popular desktop environments that run on Linux include Xfce, Mate, Cinnamon, and Ubuntu's Unity. These all come with their own unique capabilities and slightly different ways to interact within a desktop environment. Most desktops have a taskbar with a menu of available programs and a few shortcuts to frequently used programs like the browser or terminal emulator. They also usually contain some notification icons that allow the user to get a quick glance at the status of certain functions like the network, CPU utilization, or time of day. To open a program on the desktop, simply click on the shortcut in the taskbar, or click on the menu button and find the program you want to run.

The Terminal or Console

If all you want to do is browse the internet, you could stop read-ing right here—you already have the information you need to use your Raspberry Pi as a web browser. However, this book is for Makers, and Makers like to build things. So you will need to pull a lot more tools out of the Swiss Army knife that is Linux than just the desktop. To do that, you will need to get comforta-ble working in the terminal (which is also sometimes referred to as the console), shown in Figure 2-2.

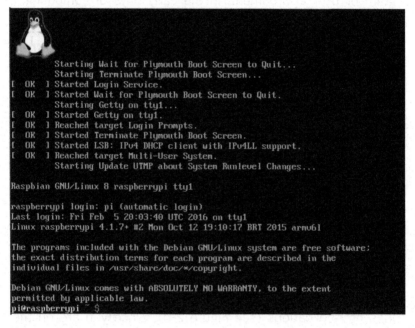

Figure 2-2. *The Linux console on a Raspberry Pi*

If you have ever followed a Raspberry Pi tutorial or how-to guide on the internet (or if you set up your Pi in Chapter 1), you were instructed to type some commands in the terminal or console.

Terminal or Console? What's the Difference?

These terms are used interchangeably these days. Strictly speaking, the terminal is a way to interface with the operating system by issuing text commands. The console, on the other hand, usually refers to a physical set of hardware (i.e., keyboard, mouse, and monitor) that provides feedback from a given program or user environment. In the early days of computing, before the advent of the desktop and after the days of punch cards, a console was the only way to interact with the computer. So one way to think about this is by saying "You can access the terminal from the console." In any case, at this point they both refer to a text-based way to interact with a computer in order to run programs.

At first, using the terminal may seem like an archaic and laborious way to get things done, but when you become proficient at using the terminal for your projects you might decide to abandon the desktop altogether (more on this later). Because the terminal was originally the only way to access a computer, and because there was no cut and paste yet, programmers designed many shortcuts in their programs that are still available today. These shortcuts make using the terminal simpler than it seems.

On the desktop, you can open a terminal window by clicking on the icon in the taskbar or by finding it in the menu under Accessories or Applications (see Figure 2-3). Since you are on the desktop, the program you open is actually *emulating* the console, so it's known as a *terminal emulator*. If you are not running the desktop, then once you boot up your Raspberry Pi or other SBC with a monitor attached, the terminal is already staring you in the face. For those not used to working on a Linux system, don't worry: you won't feel this way forever.

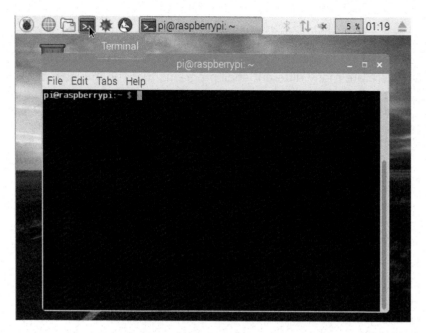

Figure 2-3. *The default Raspberry Pi terminal emulator*

The Shell in a Nutshell

The *shell* is the part of Linux that runs the terminal. The shell interprets what you type at the command prompt so the operating system knows what to do. For example, when you type the command **ls** on the command line, the shell knows where that program lives and how it should be invoked to run properly. The shell is also the mechanism that dictates how the console interface looks, and it provides a lot of the shortcuts I mentioned earlier. Think of the shell as your own personal operating system butler. I will be covering some of the most important of these shortcuts for Makers in Chapter 4.

In addition, you can create a script that the shell will run line by line. As you might expect, this is called a *shell script* and is simply a text file with some commands that get executed from top to bottom.

Try It for Yourself

At the console or terminal window, type:

```
echo Hello World!
```

Then press the Enter key and see what happens.

Now put this into a script that repeats this command 10 times by simply typing:

```
nano hello.sh
```

This tells the computer to launch **nano**, a text editor, and to edit the file *hello.sh*. If the file doesn't exist, **nano** will create it.

1 *https://en.wikipedia.org/wiki/Bash_(Unix_shell)*

When **nano** launches, type:

```
#!/bin/bash

for i in `seq 1 10`;

do

echo Hello World!

sleep 1

done
```

 Backticks

The ` characters around seq 1 10 are backticks and not single quotes. In a bash script, backticks will execute the code inside them. On an English-US keyboard, you can find the backtick key in the top-left corner next to the number 1 key. Figure 2-4 shows what this should look like when you type it in the terminal. Try typing **seq 1 10** on the command line by itself to see what it does.

Press Ctrl-X, then Y, and then press the Enter key to save the script. Now type:

```
sh hello.sh
```

and press Enter. Congratulations! You just wrote your first shell script.

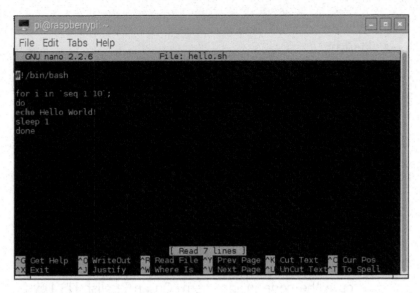

Figure 2-4. *Example bash script*

Filesystems and Structures

Something that sets Linux apart from other operating systems is its approach to filesystems. A filesystem is a way to store and organize files on a computer. As with all other Unix-based systems, almost everything you want to reference in Linux is a file. Sometimes this is intuitive. For example, when you insert a USB thumb drive into a Raspberry Pi, you can browse the files on that drive either on the desktop or in the terminal. Other times, it's not as obvious. For example, if you plug a mouse or keyboard into the Raspberry Pi, Linux creates a file to reference that device. These are special files that Linux uses to "talk" to those physical devices. Files are also used to access information about the hardware components that make up the computer itself like the CPU, memory, and other resources. If you want to find out the CPU temperature, you can just read a file.

Try It for Yourself

At the console or terminal window, type:

```
more  /sys/class/thermal/thermal_zone0/temp
```

and press the Enter key. The five-digit number that is returned represents the temperature of the CPU in millidegrees centigrade. So if your result is:

```
38470
```

then your Raspberry Pi's CPU is running at about 38 degrees centigrade.

It is important for Makers to understand the way Linux handles files if they want to use peripherals as part of their projects and if they want to programmatically get information from the operating system about certain resources. It is also important because some special files should be left alone and not deleted or overwritten. Doing so could render your system inoperable, and you might have to reinstall from scratch. Luckily, Linux has users and permissions (see "Users and Groups" on page 31) to keep you from doing too much damage by accident.

You can take a look at the files and folders of a Linux system by opening the file manager program on the desktop or by typing:

```
ls /
```

in the terminal. The forward slash by itself represents the root folder, which is the highest level in the filesystem hierarchy. On Windows, this would be similar to the *C:* drive or *C:*.

 Slash Confusion
Windows uses the backslash (\) to separate folders and filenames, while Linux uses the forward slash (/).

Linux Filesystem Structure

The basic structure of the root filesystem on a Raspberry Pi is listed here. The directories marked with a * contain sensitive information or program data and should not be deleted or changed unless you know what you are doing:

/

This is the root filesystem.

/bin*
Essential command binaries or programs are found here.

/boot*
This is a place for files that help boot up the system.

/dev*
Files that represent system devices are found here.

/etc*
This is where many configuration files for the operating system and other programs are located.

/home
Subdirectories are created here for users' home directories. If you are logged in as the default user "pi" on the Raspberry Pi, you will start in the terminal at */home/pi*.

/lib*
This is for libraries or supporting files necessary to run the programs.

/media
Removable media usually gets its own directory here when you insert it into the computer.

/mnt
This is a place to mount other filesystems. It is usually empty at first.

/opt*
Optional software or programs sometimes get installed in this directory.

/proc*
Files that represent process- or kernel-level information are kept here.

/root*
The home directory for the "root" user. It is kept separately from the other users for added security.

/run*
Current information about the running system is kept here.

/sbin*
> Here you'll find essential command binaries or programs that are "secure" and need root access to run.

/srv*
> Certain server-specific information sometimes goes here. Usually empty at first, it is sometimes used by web servers and FTP servers.

/sys*
> Information about the devices connected to the system is stored here.

/tmp
> Temporary files are often created here and then deleted with the system reboots.

/usr*
> This directory stores additional binaries and programs that are generally available to all users, although some of them require root access to run.

/var
> Variable files and data are kept here. Examples might be buffers to store printer data before it is sent to a printer, resource information, or logfiles.

Users and Groups

The ability to have multiple users with different profiles and setting is a relatively new concept for some operating systems like Windows, macOS, and Android. At the time of this writing, iOS supports multiple users only for educational use. However, since Linux is based on UNIX, and UNIX is a server operating system, it has always had support for multiple users and groups. This makes sense if you consider its roots in the time-sharing methodology (see Chapter 1). Most SBCs you use will come with a default user already set up. On Raspberry Pi, this user is called "pi." On C.H.I.P., this user is called "chip." It is always a good idea to use this default user for regular operations, but you can set up other users if you need to. I will cover those procedures in Chapter 6.

The version of Linux that comes with most SCBs will automatically log you in if you are running in desktop mode. You will need to know the password to run any advanced commands or log in when running in console mode. The default password for the "pi" user on Raspberry Pi is "raspberry." If you are using a different SBC platform, you can usually find the default user and password in the Getting Started documentation on their website. If you install Linux on a desktop or laptop PC, you will be prompted to set up a user and password during the installation process.

Linux also can combine several users together in *groups*. Using groups makes it easier to manage multiple users at the same time. Permissions granted to a group automatically apply to all users that are members of that group. Makers will most likely not need to manage Linux groups except when it is necessary to be part of a group to gain access to a device or program. I will cover this process in Chapter 6.

In addition to names, Linux also assigns a number to represent users and groups. This number is easier for the system to use than an alphabetical representation and is referred to as the *user id* (UID) and the *group id* (GID). However, what you see on the console and in other programs is almost always represented by the alphabetical name to make it easier for humans to decipher.

Every version of Linux comes with a special administrator or superuser called "root." This user has permission to do almost anything in Linux, so as you can imagine, it is not advised to log in as this user on a normal basis. In fact, most versions of Linux that run on the Raspberry Pi and similar SBCs have disabled the "root" password by default. This can be still be enabled if you really need root access, but it is not recommended. On several occasions, I have been logged in as "root" and typed the wrong command, which deleted important files or changed something I did not intend to.

Sometimes as a Maker using SBCs, it is easy just to reimage your storage and start over. However, if you have been working long and hard on a project and make a mistake while logged in as "root," it can cost you hours and perhaps days of trying to get things back to normal. This is especially true if you don't have a backup. Also, running as the "root" user on a daily basis usually leads to weaker security since we all naturally tend to make things easier for ourselves. Over time, you're bound to add backdoors and easy-to-guess passwords, making it easier for someone else to gain root access and do serious damage to your project or steal your data.

Security is usually not a big concern with most Maker projects, but if your project will be connected to the internet or in a public space, protecting its integrity is important. Several global internet outages in the fall of 2016 were traced to insecure Internet of Things (IoT) devices that had been taken over by hackers and turned into a botnet. One way to prevent your project from being used this way is to follow this basic golden rule of Linux: *don't run as the "root" user unless you have to.*

Permissions and sudo

You may be wondering how Linux makes decisions about what a user can and can't do and which files a user has access to. Perhaps you have tried to run a command or program only to get a message like "cannot remove file: permission denied." This message indicates that your user doesn't have the necessary permission level to delete a particular file. Since Linux treats most things like files, and since permissions are set on each file, it becomes important to understand what permissions do and how to use them.

Permissions are set as an attribute on each file and directory, and are arranged as a series of numbers. This series can be broken down into four groupings, which are then further broken down into read (r), write (w), and execute (x) permissions for the owner, group, and all other users, respectively. You can see an example of what this looks like in the terminal in Figure 2-5.

```
pi@raspberrypi ~ $ ls -l
total 36
drwxr-xr-x 2 pi pi 4096 Sep 24 2015 Desktop
drwxr-xr-x 5 pi pi 4096 Sep 24 2015 Documents
drwxr-xr-x 2 pi pi 4096 Sep 24 2015 Downloads
-rw-r--r-- 1 pi pi    0 Jul 13 20:31 example.txt
drwxr-xr-x 2 pi pi 4096 Sep 24 2015 Music
drwxr-xr-x 2 pi pi 4096 Sep 24 2015 Pictures
drwxr-xr-x 2 pi pi 4096 Sep 24 2015 Public
drwxrwxr-x 2 pi pi 4096 Jan 27 2015 python_games
drwxr-xr-x 2 pi pi 4096 Sep 24 2015 Templates
drwxr-xr-x 2 pi pi 4096 Sep 24 2015 Videos
pi@raspberrypi ~ $
```

Figure 2-5. *Example output of ls-l*

In Figure 2-6, I have interpreted what all these numbers and letters mean, using two of the lines from Figure 2-5.

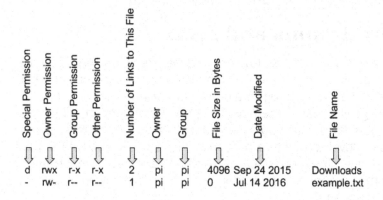

Figure 2-6. *Breakdown of permission listing on the command line*

Here are a few things we know based on this information:

1. The *Downloads* file is actually a directory, as indicated by its special permission. Other options here might be l for link, s to indicate the file should be run with owner permissions, and t to indicate that only the owner can delete or rename the file.
2. The owner of both of these files is the "pi" user.
3. The group that these files belong to is also called "pi."

4. The "pi" user has read, write, and execute permissions to the *Downloads* directory, but only read and write permissions for the *example.txt* file.
5. Members of the "pi" group have only read permission for the *example.txt* file.
6. All other users on the system (except "root") also only have read permission for the *example.txt* file.

Permissions can also be represented by a series of numbers. For example, 777 is the same as `rwx rwx rwx`. Sometimes programs will use numbers and sometimes letters to represent the permissions, so it's useful to know how to map between the two. You accomplish this by assigning a value to each of the three possible choices and then adding them together to get the permission. Here's how that works:

```
r = 4
w = 2
x = 1
```

If you want to represent `rxw`, you simply add 4+1+2 and you get 7. So a file like *example.txt*, which has permissions of `rw- r-- r--`, could also be represented as 644. We will cover how to change permissions and ownership in Chapter 6.

Try It for Yourself

In the console or terminal emulator, type:

```
ls -l
```

This will show you all the non-hidden files and directories in your current location in the filesystem. Try to figure out what the permissions are for the user and group for each file listed.

Sometimes you need to run a program or access a file that your user doesn't normally have permissions to run. You could log in as "root" to run the program, but I have already told you not to log in as "root." This presents a problem. Luckily, there is a system built into Linux that handles these special cases. That system is called **sudo**. **sudo** is short for *super user do*, and you can use it to run programs that need root permissions without logging in as "root." Using **sudo** is as easy as putting **sudo** in front of the command you want to run. sudo acts as an additional layer of security by:

- Requiring the user's password even if they are already logged in
- Expiring after a short period of time and requiring reauthentication
- Configuring sudo ability on a user-by-user basis
- Logging all sudo commands
- Eliminating the need to share the "root" password
- Rendering useless any password-cracking utilities that try to guess the "root" password if root access is disabled

Try It for Yourself

Now let's try a command we should expect to fail. In the console or terminal emulator, type:

```
apt-get update
```

Since you don't have permission to run some of the programs that apt-get uses, this command will fail and tell you:

```
E: Unable to lock the administration directory (/var/lib/
dpkg), are you root?
```

Now try again, but this time run the same command as "root":

```
sudo apt-get update
```

This time the command should update its information about available software packages if you are connected to the internet. You will learn about using apt-get in Chapter 4.

Services

In Linux, there are some programs that run when the system starts up and continue running in the background. The types of programs, and the processes that run them, are called *services*. They include things like web servers, network configuration, file sharing, remote access programs, and essential system functions. Services were developed to avoid users having to start a program from the command line every time they wanted to use it. When you boot up a Raspberry Pi, you can see a lot of text scrolling up the screen. Much of this information relates to the various services that are starting up as the system boots. You can set up almost any program to run as a service, and I will cover how to do this in Chapter 6.

Most of the time, you won't have to worry about interacting with services, and you probably shouldn't experiment with them unless you know what you are doing. For example, let's say you are running a Raspberry Pi without a monitor attached as part of a robot you are building. While you are logged in remotely, you decide to stop the networking service. Not only will your remote session immediately end, but you will have to disassemble your robot and physically connect to your Pi in order to start the service again. In the worst case, you might have to kill the power to your Pi in order to reboot it so you can connect to it again. That's never a good idea, as it could lead to data corruption on the SD card.

There is a system for managing services, which varies somewhat between different versions of Linux. On Raspberry Pis running Raspian Jessie or later, that system is called systemd. systemd (which stands for *system daemon*) controls not only services, but many other Linux resources that need to be managed, and it is the first thing that runs when the operating system boots up. For each service, there is a script that is put in a special directory when that program is installed. That script defines what happens when the service starts or stops.

Try It for Yourself

In some cases, it is necessary to restart a service or check to see if a service is actually running or not. If you want to see which services are running, you can type **systemctl** (*systemd control*) in the console or terminal emulator:

```
systemctl
```

Press the space bar to scroll down through the list one page at a time and press the Q key to exit. You can also see how long the operating system took to boot by typing:

```
systemd-analyze
```

You can see how long each service took to start by typing:

```
systemd-analyze blame
```

There are also functions for controlling each service individually. Just as a reference, I will list them here:

```
systemctl enable name . service
```

```
systemctl disable name . service
```

```
systemctl start name . service
```

```
systemctl stop name . service
```

```
systemctl restart name . service
```

```
systemctl status name . service
```

```
systemctl reload name . service
```

Processes

At any given point in time, there are multiple programs running on a Linux system. In order for them all to run at the same time without interfering with each other, there needs to be a way to keep track of them and any other supporting programs that might be needed by the original program. Linux does this with *processes*. When a program is executed, Linux creates a process to represent the work that is being done by that program. In other words, a process represents a running program. Each process receives access to system resources like the CPU, memory, and shared libraries that are needed in order for the program to work correctly. This also provides a way for the operating system to track which program is doing what in order to keep things organized and running smoothly. Processes are given a process identification number (PID) so they can be referenced more easily.

Processes have a family history. By this, I mean that there is always some program that starts another program. For example, since **systemd** (also known as the **init** process) is the first program that runs when the operating system starts, it is given the PID of 1. When **systemd** starts a service or runs a program, that program will get its own PID, but the operating system will note that its parent PID (PPID) is 1. This parent/child relationship between processes can be helpful to track down problems and issues to see where the root cause lies. I will show you how to manage processes in Chapter 6.

Try It for Yourself

To see what processes your user has started for your current console session, use the ps (process) command:

```
ps
```

Because you probably aren't running that many programs at the moment, this list will be quite small. To see a list of all the processes that are currently running, type:

```
ps -ef
```

This will give you a list of not only the processes, but which user ran them, what the PPID is, and when the process was started (see Figure 2-7).

```
UID       PID  PPID  C STIME TTY          TIME CMD
root        1    0   0 Oct17 ?        00:00:15 /sbin/init
root        2    0   0 Oct17 ?        00:00:00 [kthreadd]
root        3    2   0 Oct17 ?        00:00:05 [ksoftirqd/0]
root        5    2   0 Oct17 ?        00:00:00 [kworker/0:0H]
root        7    2   0 Oct17 ?        00:00:00 [khelper]
root        8    2   0 Oct17 ?        00:00:00 [kdevtmpfs]
root        9    2   0 Oct17 ?        00:00:00 [netns]
root       10    2   0 Oct17 ?        00:00:00 [writeback]
root       11    2   0 Oct17 ?        00:00:00 [bioset]
root       12    2   0 Oct17 ?        00:00:00 [kblockd]
root       13    2   0 Oct17 ?        00:00:00 [rpciod]
root       15    2   0 Oct17 ?        00:00:00 [kswapd0]
root       16    2   0 Oct17 ?        00:00:00 [fsnotify_mark]
root       17    2   0 Oct17 ?        00:00:00 [nfsiod]
root       18    2   0 Oct17 ?        00:00:00 [kworker/u2:1]
root       23    2   0 Oct17 ?        00:00:00 [scsi_eh_0]
root       24    2   0 Oct17 ?        00:00:00 [scsi_tmf_0]
root       25    2   0 Oct17 ?        00:00:00 [kworker/0:1H]
root       26    2   0 Oct17 ?        00:00:00 [kworker/u2:2]
root       31    2   0 Oct17 ?        00:00:00 [kpsmoused]
root       33    2   0 Oct17 ?        00:00:00 [deferwq]
root       34    2   0 Oct17 ?        00:00:00 [jbd2/sda2-8]
root       35    2   0 Oct17 ?        00:00:00 [ext4-rsv-conver]
root       67    1   0 Oct17 ?        00:00:00 /lib/systemd/systemd-udevd
root       68    1   0 Oct17 ?        00:00:04 /lib/systemd/systemd-journald
root      395    1   0 Oct17 ?        00:00:00 dhclient -v -pf /run/dhclient.et
h0.pid -lf /var/lib/dhcp/dhclient.eth0.leases eth0
root      449    1   0 Oct17 ?        00:00:01 /usr/sbin/cron -f
```

Figure 2-7. *Example output of ps -ef*

Why This Matters for Makers

Let's face it: Linux can be baffling at times, especially when you haven't used it before. Sometimes you might be able to follow a tutorial to make something work, but you might not know why it works or how to go about changing things when it doesn't work. If you've used Linux before, perhaps this chapter has answered some questions you've had. Getting stuck in the middle of a project because things aren't working is never fun and can cause days' worth of delays as you search forums looking for answers. Understanding these basic principles of Linux should give you a clue about how to diagnose the problem, as well as a good foundation for the next chapters, in which we dive into more advanced topics.

3/Using the Desktop

Although you don't need the desktop to build projects with Linux on SBCs, I will cover it here, as it will most likely be the first thing a new user will interact with when they boot up for the first time. Some people will feel more comfortable using the *graphical user interface* (GUI), so I will discuss some of the things Linux can do that aren't always obvious when you're coming from Windows or Mac.

When to Use the Desktop?

Using the Linux desktop in and of itself will probably not help you build projects with SBCs unless you need to do something graphically on the desktop, like make a game or create a GUI to interact with your project. However, there are some reasons makers might want to use the desktop anyway.

First, maybe you aren't building a project at all and just want a cheap desktop to use. I am assuming throughout this book that you're using your SBC as part of a project, like building a robot or reading sensor data. But maybe you want to use a device like the Raspberry Pi as a standalone computer to browse the web, play a small game, or write a document.

Second, you might just feel more comfortable on the desktop than you do on the command line, and there is certainly nothing wrong with that. After all, the desktop gives you an easy way out when something goes wrong. You can always reboot the system from the menu or start a second instance of the terminal emulator to poke around while your program is running in the first one.

Third, the desktop gives you access to GUI tools like a web browser, which you don't have when you are running solely in the console. This can be helpful if you don't have access to a secondary device like a laptop, desktop, or phone to look up information about your project.

When Not to Use the Desktop?

SBCs are very powerful platforms for their size. When I was a child, the idea that a fully functional desktop computer could fit in the palm of my hand was the stuff of science fiction. That being said, running anything graphical on an SBC like the Raspberry Pi takes up a lot of CPU and memory resources. If you've used a Raspberry Pi before, you may have noticed that using the browser is not the snappy experience you've come to expect on your desktop, laptop, or even on your phone. In fact, you can see the impact running programs on the desktop has on your CPU just by looking at the tiny graph in the right corner of the taskbar (see Figure 3-1).

Figure 3-1. *The CPU performance applet in the Raspberry Pi taskbar*

In this case, just launching the web browser on a Raspberry Pi sends the CPU utilization up to 100% while the program loads. If you were also running a script to control time-sensitive devices or processes, this single act could impact your project. So the first reason you may not want to run the desktop on an SBC is to conserve resources.

Another thing to keep in mind is the KISS rule. This stands for *Keep It Simple, Stupid!* The idea here is to avoid complexity unless it's absolutely required to make your program or process work. The more "moving parts" you have in a system, the greater the chance that something will go wrong. For example, if you open the hood of my 1965 Chevy pickup truck, you will only see a handful of parts. Most of the engine compartment is empty. It's easy to work on. If something goes wrong, it is easy to diagnose and fix. My 2014 Honda Odyssey is a different story. Every bit of space under the hood is taken up with some system, and many of them depend on each other to make the car run. Not only is it difficult to work on, but the complexity means that many times you need an expert to decipher what is going on when something isn't working right. Similarly, the desktop spawns processes and communication activity on the system that add a level of complexity unnecessary for most projects. You may not run into a problem by running the desktop, but why tempt fate?

For these two reasons, it's probably best to avoid using the desktop on your SBC when you don't have to. In fact, when Linux systems are used in large corporations, the desktop is almost always turned off by default to reduce the chance that something will interfere with critical business processes.

Understanding the Layout

In most respects, the layout of the desktop environment on an SBC is similar to Windows or macOS. There is usually a menu button to access programs and features. There is also an area that shows which applications are running, and a notification area that displays information like the time and date or network connectivity. The biggest difference may be that most Linux systems put the panel (also known as the taskbar) at the top of the desktop instead of the bottom. Sometimes the panel is hidden altogether and you need to right-click on the desktop to bring up the menu. See Figure 3-2 for an example of what this looks like on the Raspberry Pi.

Figure 3-2. *Breakdown of the Raspberry Pi desktop*

1. Menu button
2. Application launch bar shortcuts
3. Running applications
4. Panel applets
5. Desktop area

Connecting to the Network

Probably the first thing you will want to do if you're using the desktop is connect to the network to get internet access. If you are using a wired Ethernet connection, this should be done automatically for you. If you are using WiFi, you just need to click on the network applet icon, choose your SSID, and supply your password.

Changing the Look and Feel

There are many ways to customize the look and feel of a Linux desktop. Let's take a look at some of the more common things people like to do to personalize their desktop. I will be referencing the Raspian desktop here, but these methods should apply to other LXDEs as well.

Changing the Panel Location

To change the location of the panel from the top of the desktop to the bottom or sides, simply right-click anywhere on the panel and choose Panel Settings from the dialog box that pops up (see Figure 3-3). Then select the edge where you would like the panel to appear. You can also change the alignment, size, font, opacity, and which panel applets are loaded.

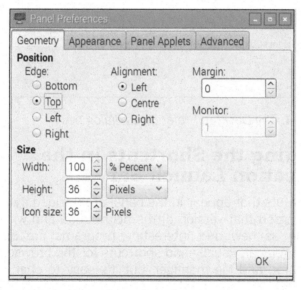

Figure 3-3. *The Panel Preferences dialog box*

Changing the Background Image

Don't care for the Raspberry Pi logo on your desktop? No problem. It is easy to change the background image to something more visually appealing. Simply right-click anywhere on the desktop and choose Desktop Preferences (see Figure 3-4). You can also change the font properties for the text that goes under the desktop shortcuts as well as some of the default icons that appear on the desktop.

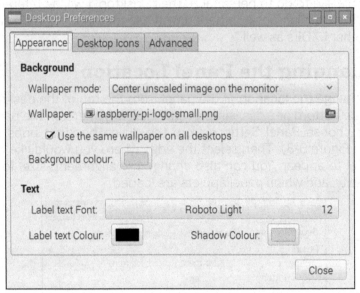

Figure 3-4. *The Desktop Preferences dialog box*

Changing the Shortcuts in the Application Launch Bar

The shortcuts that appear in the panel are decided by the people working on that version of the distribution and will change over time as new and interesting programs are added or removed. You can usually find shortcuts for the browser, terminal emulator, and file manager, but it's easy to change these shortcuts in LXDE. Just right-click on the area where the icons appear and choose Application Launch Bar Settings from the pop-up dialog box that appears (see Figure 3-5). The box on the left shows the shortcuts that are currently in the launch bar, and the box on the right shows the shortcuts available to add. You can add or remove application shortcuts by clicking on a shortcut and using the Add and Remove buttons. You can also change the order in which the shortcuts appear in the launch bar using the Up and and Down buttons.

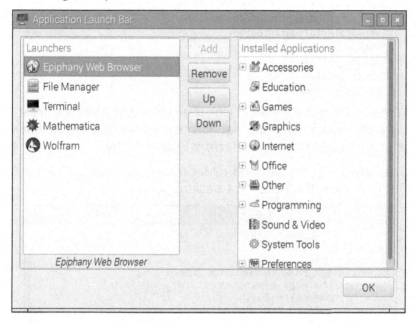

Figure 3-5. *The Application Launch Bar dialog box*

In other distributions, you can browse through the menu, right-click on the application you want to add, and choose "Add to panel," or sometimes you can click and drag the icon to the panel.

Creating a Desktop Shortcut

It is also very straightforward to add an LXDE desktop shortcut to an application that is already in the menu. Just navigate through the menu to the application in question, right-click on it, and choose "Add to desktop."

The process of creating a desktop shortcut to launch an application that's not in the menu or for a script that you wrote is a little more cumbersome. Open the built-in text editor by choosing Accessories→Text Editor. Then type in the following information about the shortcut you want to create:

```
[Desktop Entry]
Name=Some Name
Comment=Click here to run this thing
Icon=/usr/share/pixmaps/openbox.xpm
Exec=/path/to/your/program
Type=Application
Encoding=UTF-8
Terminal=false
Categories=None;
```

When you're done, it should look like Figure 3-6. If you are creating a shortcut for a script that normally runs in a terminal emulator, be sure to change the Terminal value to true.

Save the file in the *Desktop* folder under your home directory. Be sure to give the filename a *.desktop* suffix.

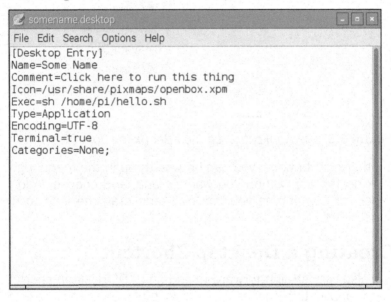

Figure 3-6. *An example of a desktop shortcut configuration file*

In other Linux desktop environments, you can right-click on the desktop and select "Create launcher" or "Add shortcut."

Try It for Yourself

Try to create a desktop shortcut that runs the *hello.sh* shell script that you created in Chapter 2. See Figure 3-6 if you need some hints on what the file should look like.

Why This Matters for Makers

Most Makers who are new to Linux will naturally start by using the desktop. This is a safe and somewhat familiar place to try things out and look around. Later, you may want to disable the desktop and save precious resources on your SBC. We will cover how to do that in Chapter 5.

4/Command-Line Basics

In this chapter, I will show you some basic ways to use the command line that work whether you are running a terminal emulator on the desktop or have a keyboard and monitor plugged directly into your Raspberry Pi. Over the years, many tricks and shortcuts have been built into the Linux operating system to make using the command line quick and easy. Let's take a look at some of the first things every Maker should know how to do when they land on the command line.

Understanding the Prompt

The prompt is the indicator on the command line that shows where you are on the screen. In other words, the computer is "prompting" you to type something at a specific place. The prompt won't appear until you are logged in to the system. When you type something into the terminal, that something will show up at the prompt. Figure 4-1 shows what the prompt looks like by default on the Raspberry Pi.

Figure 4-1. *The Linux bash prompt on Raspberry Pi*

The default prompt on the Raspberry Pi and most other Linux systems is configured to give us several pieces of useful information at a glance. Let's change to the *Downloads* directory and break the prompt down into its component parts (see Figure 4-2):

```
cd Downloads
```

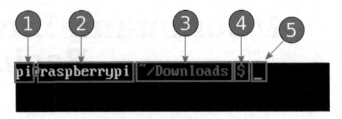

Figure 4-2. *Breakdown of the prompt*

1. Your username.
2. The hostname of the Raspberry Pi.
3. Your current location in the filesystem.
4. The symbol that indicates this is the prompt. If you are "root," this changes to a #.
5. A cursor to show you where text will be placed as you type. You can move this around with the arrow keys to adjust its position.

Notice that the prompt is green and blue. This is just to help distinguish between different parts of the prompt to avoid confusion. When you press Enter, whatever you type and any output of commands that you run will be scrolled up the screen, moving the prompt down to the next available line. If you are on the desktop using a terminal emulator, you can scroll back and forth using the mouse wheel or the scroll bar on the side of the window. If you are not on the desktop, you can scroll back and forth by holding down the Shift key and pressing the Page Up or Page Down keys. The Shift+Page Up/Down function works in almost every version of Linux. Also, while typing a command, you can use the arrow keys or the Home and End buttons to move your cursor where you want it.

Try It for Yourself

Most Linux terminals only display 30 to 40 lines of text before information starts scrolling off the top, depending on the size of your screen or window. Inevitably, there will be some output that you want to see that scrolls off the screen before you can get to it. For example, let's run a command that I know will generate a lot of output. *Display message* (dmesg) is a command that prints all the system messages that come from the Linux kernel, going back to when the system was booted. Type the following on the console or terminal emulator and press the Enter key:

```
dmesg
```

Looking at the output, there is no way to read all of the information as it scrolls by so fast. Use the Shift+Page Up or Shift+Page Down key combination to scroll up and down through the output one page at a time.

Orienting Yourself in the Filesystem

When using Linux, it's important to know what directory you are located in before you run a command. It's also useful to know how to maneuver from one location to another and get information about the files you are working with. Before you do anything else, you need to learn these basic techniques for working on the Linux command line.

Where Am I?: pwd

Most of the time, you can figure out which directory you are in just by looking at the prompt. But some systems don't offer such a detailed prompt, and other times you might just want to make sure. You can find out where you are with the command pwd, which stands for *print working directory*. This will print one line that shows you exactly where you are in the directory structure (see Figure 4-3). Remember, the highest-level location in the directory structure is /.

Figure 4-3. *An example of the output of the pwd command*

So, from Figure 4-3 we know that we are located in a directory called *pi*, which is a subdirectory of *home*, which is a subdirectory of /. This is the "pi" user's home directory. By default, when you log on to the system or open a terminal emulator, you start in your home directory. Anything that you do on the command line (create a file, delete a file, run a script or program) will be done based on where in the filesystem you are currently located. You may be wondering why the prompt in Figure 4-3 isn't showing */home/pi* as the current location; instead, it's just showing ~. That's because ~ is an alias for the current user's home directory. You can find out more about aliases in Chapter 6.

Changing the Working Directory: cd

You change location from one directory to another in Linux by using the command **cd**, which stands for *change directory*. To get to a different place in the directory structure, just type cd, then a space followed by the path to your destination. Keep in mind that everything in Linux is case-sensitive. For example, typing:

```
cd /home/pi/Downloads
```

will move you from */home/pi* into */home/pi/Downloads*. You can think of it as moving from room to room in a big house. Even though the house is very large, you can move wherever you want as long as you know the path to that room. In this case, your front door or starting location is always */*. You don't always have to operate linearly when you change to a new location. You can jump wherever you want. For example, typing:

```
cd /var/log
```

will move you from wherever you are to */var/log/*, the location where logfiles are kept. In this way, the cd command also acts like a teleporter, instantly moving you wherever you want to go in the house. When you compare this to a graphical file manager where you may have to double-click your way back and forth through the directory structure, running a single command on the command line is much quicker and easier, albeit not quite as intuitive.

Typing in the full path to the directory you want to change to can be cumbersome. This full path that starts with / and ends with the desired destination (i.e., */home/pi/Downloads*) is called an *absolute path*. You can also use *relative paths* that will save you some time. For example, previously when we wanted to move from */home/pi* into */home/pi/Downloads* we typed:

```
cd /home/pi/Downloads
```

However, you can do the same thing by just typing:

```
cd Downloads
```

This works because *Downloads* is a subdirectory of */home/pi* and we are already located at */home/pi*. Notice there is no leading / in front of *Downloads*. That's because we don't want to move to a top-level subdirectory of /. Instead, we want to move to a subdirectory that is relative to our current location.

Not only can you descend into subdirectories by using relative paths, but you can change to the parent directory of your current location as well by using the .. alias. In Linux, . usually refers to your current location, whereas .. refers to the location one level up from where you are. So if you are located at */home/pi/Downloads* and want to move up one level to the */home/pi* directory, you can simply type:

```
cd ..
```

If instead you wanted to move two levels up to the */home* directory, you can type:

```
cd ../..
```

Remember I said that ~ is an alias or shortcut for a user's home directory. You can always get back to your home directory just by typing:

```
cd ~
```

You can also get to your home directory just by typing **cd** with no arguments:

```
cd
```

Want to get to your *Downloads* subdirectory in your home directory from anywhere in the system? Just type:

```
cd ~/Downloads
```

This is equivalent to typing:

```
cd /home/pi/Downloads
```

Just replace */home/pi* with ~ whenever you want to refer to that location.

Printing Out the Contents of a Directory: ls

When you get to the directory that's your destination, you will certainly want to look around. You can use the command ls, which stands for *list*. This command prints out an alphabetically sorted list of files and directories in your current location (see Figure 4-4).

```
pi@raspberrypi ~ $ ls
Desktop    Downloads    hello.sh  Pictures  python_games  Videos
Documents  example.txt  Music     Public    Templates
pi@raspberrypi ~ $
```

Figure 4-4. *An example of running the ls command*

Since we are using a prompt with colors enabled, we can tell the difference between the directories and files just by looking at the colors. In this case, the blue text represents directories and the gray text represents files. However, we can get more detailed information simply by adding the -l option to the end of the command (see Figure 4-5).

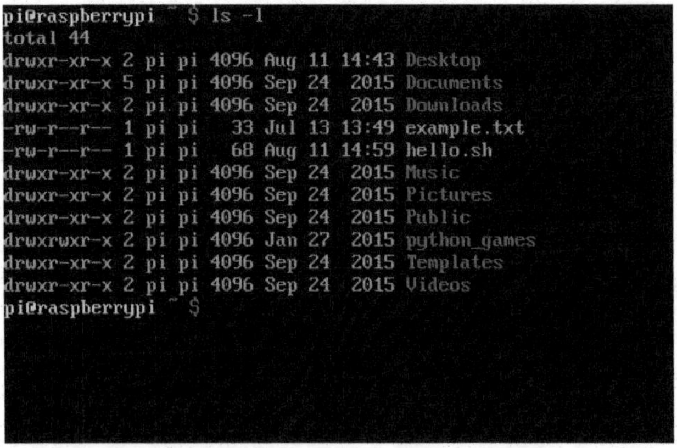

```
pi@raspberrypi ~ $ ls -l
total 44
drwxr-xr-x 2 pi pi 4096 Aug 11 14:43 Desktop
drwxr-xr-x 5 pi pi 4096 Sep 24  2015 Documents
drwxr-xr-x 2 pi pi 4096 Sep 24  2015 Downloads
-rw-r--r-- 1 pi pi   33 Jul 13 13:49 example.txt
-rw-r--r-- 1 pi pi   68 Aug 11 14:59 hello.sh
drwxr-xr-x 2 pi pi 4096 Sep 24  2015 Music
drwxr-xr-x 2 pi pi 4096 Sep 24  2015 Pictures
drwxr-xr-x 2 pi pi 4096 Sep 24  2015 Public
drwxrwxr-x 2 pi pi 4096 Jan 27  2015 python_games
drwxr-xr-x 2 pi pi 4096 Sep 24  2015 Templates
drwxr-xr-x 2 pi pi 4096 Sep 24  2015 Videos
pi@raspberrypi ~ $
```

Figure 4-5. *An example of running the ls -l command*

We covered what all this information means in Chapter 2. However, there are actually more files and directories in this location than are listed here. This is because Linux sometimes hides files to make them slightly more difficult to interact with or so they don't clutter things up when you're trying to find something. In Linux, hidden files and directories have a filename that starts with a . character. You can display these hidden files by adding the option -a to the ls command (see Figure 4-6).

Chain Options Together

Many times, you can chain options together on the command line. For example, instead of typing ls -l -a, you can simply type ls -la.

```
pi@raspberrypi ~ $ ls -la
total 116
drwxr-xr-x 20 pi    pi    4096 Sep  2 12:20 .
drwxr-xr-x  3 root  root  4096 Sep 24  2015 ..
-rw-------  1 pi    pi    2279 Sep  9 12:51 .bash_history
-rw-r--r--  1 pi    pi    3243 Sep 24  2015 .bashrc
drwxr-xr-x  6 pi    pi    4096 Feb  3  2016 .cache
drwx------ 11 pi    pi    4096 Aug 11 17:08 .config
drwx------  3 pi    pi    4096 Sep 24  2015 .dbus
drwxr-xr-x  2 pi    pi    4096 Aug 11 14:43 Desktop
drwxr-xr-x  5 pi    pi    4096 Sep 24  2015 Documents
drwxr-xr-x  2 pi    pi    4096 Sep 24  2015 Downloads
-rw-r--r--  1 pi    pi      33 Jul 13 13:49 example.txt
drwxr-xr-x  2 pi    pi    4096 Feb  3  2016 .gstreamer-0.10
-rw-r--r--  1 pi    pi      68 Aug 11 14:59 hello.sh
drwx------  3 pi    pi    4096 Aug 11 12:49 .local
drwxr-xr-x  9 pi    pi    4096 Aug 11 12:47 .Mathematica
drwxr-xr-x  2 pi    pi    4096 Sep 24  2015 Music
drwxr-xr-x  2 pi    pi    4096 Sep 24  2015 Pictures
-rw-r--r--  1 pi    pi     675 Sep 24  2015 .profile
drwxr-xr-x  2 pi    pi    4096 Sep 24  2015 Public
drwxrwxr-x  2 pi    pi    4096 Jan 27  2015 python_games
drwxr-xr-x  2 pi    pi    4096 Sep 24  2015 Templates
drwxr-xr-x  3 pi    pi    4096 Sep 24  2015 .themes
drwx------  4 pi    pi    4096 Aug 11 14:09 .thumbnails
drwxr-xr-x  2 pi    pi    4096 Sep 24  2015 Videos
drwxr-xr-x  9 pi    pi    4096 Aug 11 12:47 .WolframEngine
-rw-------  1 pi    pi     105 Aug 31 00:47 .Xauthority
-rw-------  1 pi    pi    5295 Sep  2 12:20 .xsession-errors
pi@raspberrypi ~ $
```

Figure 4-6. *The output of ls -la*

The output of ls -la gives us both detailed output as well as showing all the hidden files and directories in our current location. There are many other useful options for the ls command. Here are some of them:

`-a`

List all files including hidden files starting with "."

`-l`

List with long format and show permissions

`-lh`

List long format with human-readable file sizes

`-s`

List with file size listed first

`-r`

List in reverse order

`-R`

List recursively directory tree

`-S`

Sort by file size

`-t`

Sort by time and date

`-X`

Sort by extension name

You may wonder what the "total 116" means in the first line returned in Figure 4-6. That represents the total number of disk blocks used by all the files listed by the `ls -l` command. Not very useful, but now you know.

Creating New Files and Directories: mkdir and touch

Sometimes you want to organize your files into directories other than the ones that already exist. To make a new directory, you can use the command `mkdir`, which stands for *make directory*. To use the command, simply type:

```
mkdir mynewdirectory
```

To make a series of subdirectories under your new one, go into that subdirectory with the `cd` command and then type:

```
mkdir subdirectory1
cd subdirectory1
mkdir subdirectory2
cd subdirectory2
mkdir subdirectory3
```

Or you could make them all at the same time by using the -p option like this:

```
mkdir -p subdirectory1/subdirectory2/subdirectory3
```

You can also create files in your current location. This is done in several ways. Most of the time, files are created by the applications you are using. In Chapter 2, when you created your first shell script, you used the program nano to create the file *hello.sh*. This file was created when you saved it before exiting nano. However, you can also create empty files with the command touch. These files have no data associated with them except for the filename itself, permissions, and ownership. To create an empty file, just type **touch** and the name of the file you want to create like this:

```
touch emptyfile
```

There are a couple of reasons why you might want to create an empty file. First, you might want to double-check that you have write permissions in your current location. If you can create a file with touch, you know you have write permissions. Second, you might want to create some placeholder files if you are setting up a file structure for a complex project like a website or a program that will use multiple config files and logfiles.

Moving and Deleting Files: cp, mv, and rm

Quite often, you will find you need to move files from one location to another or delete them altogether. You can make a copy of a file by using the command cp, which stands for *copy*. Likewise, you can move a file using the command mv, which of course stands for *move*. The copy and mv commands work in similar ways. To copy a file, type **cp** followed by the file you want to copy, followed by the destination path. For example:

```
cp /home/pi/hello.sh /home/pi/Downloads/
```

This will copy the file *hello.sh* into the *Downloads* subdirectory of your home directory. Of course, if you are already located in your home directory, you could accomplish the same thing with relative paths like this:

```
cp hello.sh Downloads
```

If instead you wanted to move the *hello.sh* file into the *Documents* folder, you would type:

```
mv hello.sh Documents
```

To delete a file, use the remove command (rm). *Be careful with this command!* There is no recycle bin or trash can in Linux. Once you delete a file using the command line, it's gone for good. Though there are utilities that can sometimes recover files, they are for advanced users and don't always work. If you're sure you want to delete a file, it's simply a matter of typing **rm** followed by the filename:

```
rm hello.sh
```

If you want to delete a whole directory including all subdirectories, you can use the -R option like this:

```
rm -R directory
```

Again, this is a powerful but dangerous operation, so use it with caution. Occasionally, online trolls will try to prank new Linux users into running the command sudo rm -fR /. This is *never* a good idea, as it will wipe out your whole filesystem.

Try It for Yourself

Practice navigating around the filesystem by using the command line. Let's start in our home directory. To make sure you are there, simply type:

```
cd
```

Now let's create a subdirectory called *Sub*:

```
mkdir Sub
```

List the files in reverse time order to make sure that the directory was created:

```
ls -ltr
```

Now change into that subdirectory:

```
cd Sub
```

Make sure you are now located in that subdirectory:

```
pwd
```

Now create a test file:

```
touch testfile
```

Verify that the file is there:

```
ls -l
```

Make a copy of the file:

```
cp testfile testfile2
```

Verify that the copy was made:

```
ls -l
```

Now delete the original file:

```
rm testfile
```

Change back to your home directory:

```
cd ../
```

Now delete the whole subdirectory including the copy of the file we made:

```
rm -R Sub
```

Verify that the subdirectory is gone:

```
ls -l
```

Figure 4-7 shows a visual example of what this should look like.

```
pi@raspberrypi   $ cd
pi@raspberrypi ~ $ mkdir Sub
pi@raspberrypi ~ $ ls -ltr
total 48
drwxrwxr-x 2 pi pi 4096 Jan 27  2015 python_games
drwxr-xr-x 5 pi pi 4096 Sep 24  2015 Documents
drwxr-xr-x 2 pi pi 4096 Sep 24  2015 Downloads
drwxr-xr-x 2 pi pi 4096 Sep 24  2015 Videos
drwxr-xr-x 2 pi pi 4096 Sep 24  2015 Templates
drwxr-xr-x 2 pi pi 4096 Sep 24  2015 Public
drwxr-xr-x 2 pi pi 4096 Sep 24  2015 Pictures
drwxr-xr-x 2 pi pi 4096 Sep 24  2015 Music
-rw-r--r-- 1 pi pi   33 Jul 13 13:49 example.txt
drwxr-xr-x 2 pi pi 4096 Aug 11 14:43 Desktop
-rw-r--r-- 1 pi pi   68 Aug 11 14:59 hello.sh
drwxr-xr-x 2 pi pi 4096 Sep 18 21:31 Sub
pi@raspberrypi ~ $ cd Sub
pi@raspberrypi ~/Sub $ pwd
/home/pi/Sub
pi@raspberrypi ~/Sub $ touch testfile
pi@raspberrypi ~/Sub $ ls -l
total 0
-rw-r--r-- 1 pi pi 0 Sep 18 21:31 testfile
pi@raspberrypi ~/Sub $ cp testfile testfile2
pi@raspberrypi ~/Sub $ ls -l
total 0
-rw-r--r-- 1 pi pi 0 Sep 18 21:31 testfile
-rw-r--r-- 1 pi pi 0 Sep 18 21:31 testfile2
pi@raspberrypi ~/Sub $ rm testfile
pi@raspberrypi ~/Sub $ cd ..
pi@raspberrypi ~ $ rm -R Sub
pi@raspberrypi ~ $ ls -l
total 44
drwxr-xr-x 2 pi pi 4096 Aug 11 14:43 Desktop
drwxr-xr-x 5 pi pi 4096 Sep 24  2015 Documents
drwxr-xr-x 2 pi pi 4096 Sep 24  2015 Downloads
-rw-r--r-- 1 pi pi   33 Jul 13 13:49 example.txt
-rw-r--r-- 1 pi pi   68 Aug 11 14:59 hello.sh
drwxr-xr-x 2 pi pi 4096 Sep 24  2015 Music
drwxr-xr-x 2 pi pi 4096 Sep 24  2015 Pictures
drwxr-xr-x 2 pi pi 4096 Sep 24  2015 Public
drwxrwxr-x 2 pi pi 4096 Jan 27  2015 python_games
drwxr-xr-x 2 pi pi 4096 Sep 24  2015 Templates
drwxr-xr-x 2 pi pi 4096 Sep 24  2015 Videos
pi@raspberrypi ~ $
```

Figure 4-7. *Typical output of the previously listed commands*

Get Help with a Command: help, man, and info

There are tens of thousands of Linux commands and programs that can be run from the command line, and most of them have several options. There is no way we can cover them all in this book. Luckily, there are easy resources that can help you figure out how to use a command you aren't familiar with yet. These resources are available even if you are offline, because they are either part of the program itself or automatically installed when the program is installed.

You can usually access the first of these by simply running the command itself with a help option added. This option can take multiple forms depending on the person who wrote the program, but usually it is one of these four:

- `--help`
- `--h`
- `-help`
- `-h`

Most programmers who write utilities like `ls` or `mkdir` include some basic helpful information like usage and common options, which print out when you use the help option. In fact, most utilities will even tell you what the help option is if you don't use it correctly the first time (see Figure 4-8).

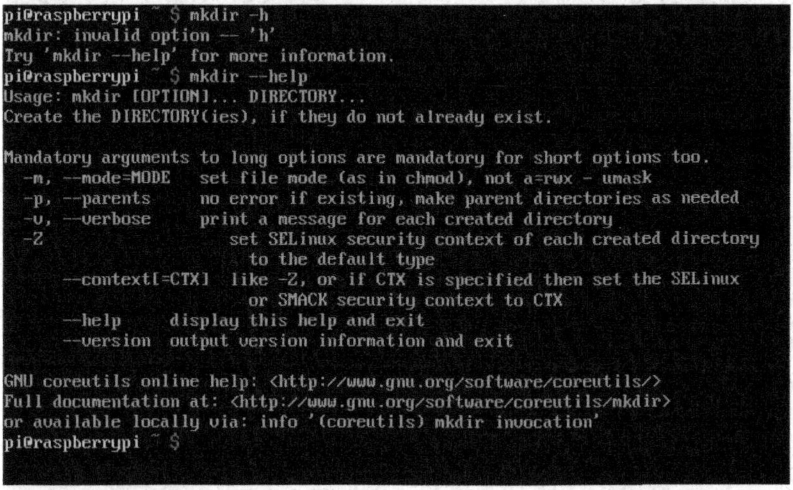

```
pi@raspberrypi ~ $ mkdir -h
mkdir: invalid option -- 'h'
Try 'mkdir --help' for more information.
pi@raspberrypi ~ $ mkdir --help
Usage: mkdir [OPTION]... DIRECTORY...
Create the DIRECTORY(ies), if they do not already exist.

Mandatory arguments to long options are mandatory for short options too.
  -m, --mode=MODE   set file mode (as in chmod), not a=rwx - umask
  -p, --parents     no error if existing, make parent directories as needed
  -v, --verbose     print a message for each created directory
  -Z                set SELinux security context of each created directory
                      to the default type
      --context[=CTX] like -Z, or if CTX is specified then set the SELinux
                      or SMACK security context to CTX
      --help        display this help and exit
      --version     output version information and exit

GNU coreutils online help: <http://www.gnu.org/software/coreutils/>
Full documentation at: <http://www.gnu.org/software/coreutils/mkdir>
or available locally via: info '(coreutils) mkdir invocation'
pi@raspberrypi ~ $
```

Figure 4-8. *Printing out the help information screen for mkdir*

The second resource available in Linux to help you learn how to use a command is man, which stands for *manual*. Yes, there is actually a built-in manual that comes with every installation of Linux. As new programs or commands are added and installed, pages are added to the manual that explain how to use them. Consider man to be your own offline encyclopedia of Linux commands and utilities. man started in the early days of Linux as a way to document the operating system and associated programs. It has since become a standard for people writing Linux utilities to include a section in the manual.

When you look up information about a specific command or program using man, the content that is displayed is called that program's manpage. If a program comes as part of Linux or is distributed as part of an official Linux repository, it probably has a manpage. A manpage is typically more detailed than the output of using a command's help option. Accessing a manpage is a very simple process. Simply type **man** followed by the command you want to learn about. For example, Figure 4-9 shows the output of typing man mkdir.

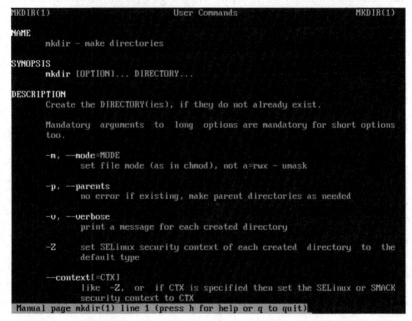

Figure 4-9. *The manpage for the mkdir command*

Once you open the manpage, you can navigate through the document to find the information you're looking for. Using the following commands will make that a little easier:

h

Help screen

Down arrow or Enter key
    Move down one line

Up arrow
    Move up one line

f, *space bar, or Page Down key*
    Page down one screen at a time

b *or Page Up key*
    Page up one screen at a time

G
    Jump to the last line in the file

g
    Jump to the first line in the file

/pattern
    Search for pattern

q
    Quit

Manpages have become standardized over the years and always contain at least the following sections:

Name
    The name of the command or function, followed by a one-line description of what it does.

Synopsis
    A formal description of how to run the command and what command-line options it takes.

Description
    A description of how the command functions.

Examples
    Usage examples.

See Also
    Related commands or functions.

You will notice in Figure 4-9 that the title is *MKDIR(1)*. The (1) refers to which section of the manual this page is located in. In this case, it is *User Commands*. Sometimes programs will have information in more than one section. When they do, it will be called out under *SEE ALSO*. Here is a breakdown of the sections so you will understand what the numbers refer to:

1. User Commands
2. System Calls
3. C Library Functions
4. Devices and Special Files
5. File Formats and Conventions
6. Games et al.
7. Miscellanea
8. System Administration tools and Daemons

Distributions customize the manual section to their specific needs and often include additional sections. If you notice that the same command is included in a different section and you would like to look at it, simply add the section number before the command name:

`man printf`

This will give you information on how to use `printf` (a command used to format text output) on the command line. However:

`man 3 printf`

will show you the manpage explaining how the `printf` command can be used in a C program.

The third way to get information in Linux is the `info` utility. Similar to `man`, to look up information in an info file, simply type **info** followed by the command you want to learn about. Figure 4-10 shows the output of typing `info mkdir`.

```
File: coreutils.info,  Node: mkdir invocation,  Next: mkfifo invocation,  Prev:\
ln invocation,  Up: Special file types

12.3 ■mkdir■: Make directories
=================================

■mkdir■ creates directories with the specified names.  Synopsis:

    mkdir [OPTION]■ NAME■

    ■mkdir■ creates each directory NAME in the order given.  It reports
an error if NAME already exists, unless the ■-p■ option is given and
NAME is a directory.

    The program accepts the following options.  Also see ★note Common
options::.

■-m MODE■
■--mode=MODE■
    Set the file permission bits of created directories to MODE, which
    uses the same syntax as in ■chmod■ and uses ■a=rwx■ (read, write
    and execute allowed for everyone) for the point of the departure.
    ★Note File permissions::.

    Normally the directory has the desired file mode bits at the moment
    it is created.  As a GNU extension, MODE may also mention special
    mode bits, but in this case there may be a temporary window during
    which the directory exists but its special mode bits are incorrect.
--zz-Info: (coreutils.info.gz)mkdir invocation, 66 lines --Top------------
Welcome to Info version 5.2.  Type h for help, m for menu item.
```

Figure 4-10. *The info page for the mkdir command*

From the top of the screen, you can tell that we are browsing the file called *coreutils.info* and have jumped to the node (section) on mkdir. You can also tell that the next node is about mkfifo and the previous node is about ln. To navigate when browsing an info file, you can use the following keys:

h
> Access the help window.

x
> Close the help window.

q
> Quit info altogether.

H
> Invoke the info tutorial.

Up
> Move up one line.

Down
> Move down one line.

Delete
> Scroll backward one screenful.

Space bar
> Scroll forward one screenful.

Home
> Go to the beginning of this node.

End
> Go to the end of this node.

Tab
> Skip to the next hypertext link.

Enter
> Follow the hypertext link under the cursor.

l
> Go back to the last node seen in this window.

[
> Go to the previous node in the document.

]
> Go to the next node in the document.

p
> Go to the previous node on this level.

n
> Go to the next node on this level.

u
> Go up one level.

t
> Go to the top node of this document.

d
> Go to the main directory node.

s
> Search forward for a specified string.

{

Search for previous occurrence.

}

Search for next occurrence.

i

Search for a specified string in the index, and select the node referenced by the first entry found.

Those are three ways to get information about a command from within Linux itself. Of course, your best resource these days outside of the operating system is just to search on the internet using your favorite search engine. Also, once you are ready to do a deep dive on something like running a web server or advanced programming, you could buy a great book like this one.

Try It for Yourself

Let's practice learning about commands using the command line, starting with pwd:

```
man pwd
```

Scroll through the manpage using the various directional keys, and when satisfied, press the Q key to quit. Now do the same for the other commands we've learned in this chapter:

```
man cd
man ls
man mkdir
man touch
man cp
man mv
man rm
```

Eliminate Some Typing

If you're like me, your spelling and typing abilities may be lacking. Too many times, I have spent 20 or 30 seconds typing a long command with lots of options only to find out after I hit the Enter key that I had typed something wrong and needed to start over from the beginning. Not only that, but it can be hard to remember the exact command you use to perform a certain task from day to day. Luckily, the Linux shell has some tools built in that can help with both of these problems.

Auto-Complete a Command: Tab

You can use the auto-complete feature of the shell by simply pressing the Tab key on the keyboard. This will auto-complete a command that has been partially typed. It will also auto-complete a filename based on the context of what you are typing.

For example, if you type **tou** and press the Tab key, the shell will fill in the rest of the missing letters to make **touch**. If there are multiple options that start with the letters you've entered, the first time you press Tab, nothing will happen. If you press it again, however, the shell will display a list of all possible commands or filenames that start with the letters you entered. So if you type mkd and press Tab twice, you will be presented with two options for commands that start with mkd: `mkdir` and `mkdosfs` (see Figure 4-11).

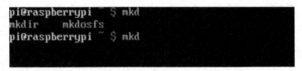

Figure 4-11. *Using the Tab key for auto-completion of commands*

If you continue to add more characters and then press Tab, you will eventually rule out all the other options and the shell will complete the rest of the command or filename when there is only one choice left. This auto-complete feature is a real time saver with bigger commands and long filenames. It also eliminates spelling errors when you haven't used a command very often.

To Tab or Not to Tab

When you're using the default bash shell in Linux, Tab does not know about the available options for a command; it only knows the name of the command and any associated filenames that might be used as part of a command.

Search for a Previous Command: Up, Ctrl-R

Linux keeps a history of all the things you type into the command line. A simple way to review the commands you've typed is to use the up arrow key to scroll back through each command starting with the most recent. When the command you're looking for is far back in your history, you can search for it by typing Ctrl-R on the command line followed by some characters. For example, if you wanted to search for the last time you used nano to edit a file, you would type Ctrl-R followed by nano. It doesn't matter if there is already some information entered at the cursor when you press Ctrl-R. That text won't be used for the search—only what you type after you press Ctrl-R. Notice that the prompt changes to (reverse-i-search) followed by the letters you entered when doing this type of search through your command history. If you press one of the arrow keys, Home, End, or Tab, you will finish the search and be able to edit the command that you looked up. You can also continue to search through your history by pressing Ctrl-R multiple times before you exit out of the search (see Figure 4-12).

Figure 4-12. *Using Ctrl-R to search the command history*

Try It for Yourself

Change to your home directory and create a file by typing:

```
cd
```

```
tou <Tab> file1
```

When you press Tab, it should complete the name of the **touch** command. Now change to your *Downloads* directory by typing:

```
cd D <Tab> <Tab>
```

You should see something similar to Figure 4-13.

Figure 4-13. *Using Tab to auto-complete a directory or filename*

Add the letters **ow** and press Tab again to auto-complete the path you want and press Enter.

Now let's create our second file by using the command history. Press Ctrl-R and then type **tou** (see Figure 4-14).

Figure 4-14. *Using command history to look up a previous command*

Press the End key and change *file1* to *file2*. Press Enter to complete the task. Now you've created two files: one in your home directory and one in the *Downloads* directory. You have also saved a lot of typing in the process.

Connecting to the Network via the Command Line

To be honest, configuring your device to connect to a wireless network via the command line is pretty complicated, especially when compared to the point-and-click ease of using the desktop. It is important to know the basics of how to do this, just in case you ever have to. Many times I have taken my projects to a Maker Faire or interview only to find that my network was still configured to connect to my home router. If you'll be taking your network-connected projects on the road, you will need to know how to configure the network so your project can work properly.

Easy Portability

When traveling, consider connecting your SBC's WiFi to your tethered phone or other portable hotspot device that creates its own WiFi network. This way, your project will connect and have access to the internet wherever you go, as long as you have cellular service.

The Interfaces File

To configure the network from the command line, you use a configuration file named *interfaces*. Its full path is */etc/network/interfaces*. This file is read when the system boots and whenever a networking interface is enabled. You can use nano or another text editor to make changes to this file. As of this writing, the configuration of the interfaces file looks like Figure 4-15.

```
# interfaces(5) file used by ifup(8) and ifdown(8)

# Please note that this file is written to be used with dhcpcd
# For static IP, consult /etc/dhcpcd.conf and 'man dhcpcd.conf'

# Include files from /etc/network/interfaces.d:
source-directory /etc/network/interfaces.d

auto lo
iface lo inet loopback

iface eth0 inet manual

allow-hotplug wlan0
iface wlan0 inet manual
    wpa-conf /etc/wpa_supplicant/wpa_supplicant.conf

allow-hotplug wlan1
iface wlan1 inet manual
    wpa-conf /etc/wpa_supplicant/wpa_supplicant.conf
```

Figure 4-15. *The default interfaces file on Raspberry Pi*

Let's break this down a bit to better understand what's going on here before we start changing things around.

- Lines that start with the # symbol are comments and are ignored.
- source-directory refers to a location where other configuration files might be stored. By default, this directory is empty and exists just in case you make changes later.
- lo, eth0, wlan0, and wlan1 refer to network interfaces on the SBC. These interfaces either exist currently or might exist at some point in the future. For example, wlan0 or wlan1 might be created when you plug a WiFi adapter into a USB port.
- iface starts the configuration section for a particular interface. All the lines underneath the iface line relate to the configuration of this interface.
- inet specifies that we will be configuring this interface for TCP/IP communication.
- loopback, manual, dhcp, static, and a few others refer to the way configuration data will be assigned to this interface.
- allow-hotplug will automatically attempt to configure the interface when the system detects that it has been connected.
- wpa-conf refers to the location of a separate, securely stored WiFi configuration file.

Wired Ethernet

By default, if your local network is set up to automatically assign an IP address with dhcp, there is nothing more you should have to do to the configuration file to get a wired Ethernet connection working. When you plug in an Ethernet cable to your Raspberry Pi, it should connect to the network, grab the appropriate configuration information, and be ready to browse the internet in just a few seconds.

Static IP Address

Sometimes you may want to manually assign your own IP address and other configuration information. Figure 4-16 shows how you would change the *interfaces* configuration file to assign static IP address information.

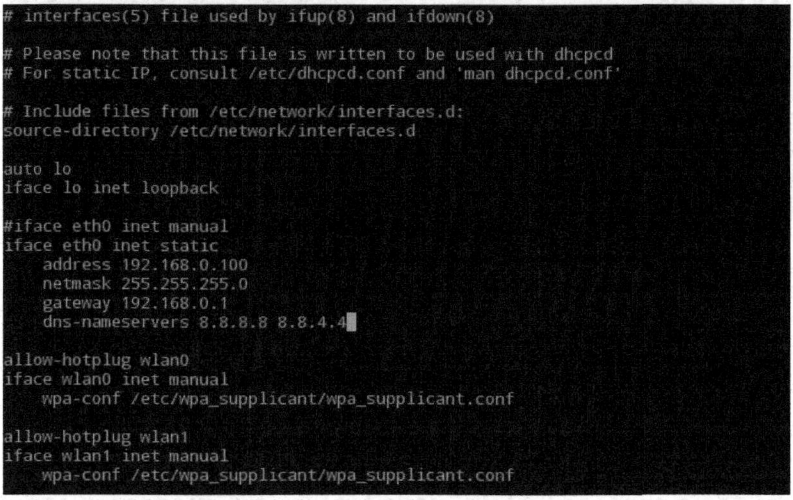

```
# interfaces(5) file used by ifup(8) and ifdown(8)

# Please note that this file is written to be used with dhcpcd
# For static IP, consult /etc/dhcpcd.conf and 'man dhcpcd.conf'

# Include files from /etc/network/interfaces.d:
source-directory /etc/network/interfaces.d

auto lo
iface lo inet loopback

#iface eth0 inet manual
iface eth0 inet static
    address 192.168.0.100
    netmask 255.255.255.0
    gateway 192.168.0.1
    dns-nameservers 8.8.8.8 8.8.4.4█

allow-hotplug wlan0
iface wlan0 inet manual
    wpa-conf /etc/wpa_supplicant/wpa_supplicant.conf

allow-hotplug wlan1
iface wlan1 inet manual
    wpa-conf /etc/wpa_supplicant/wpa_supplicant.conf
```

Figure 4-16. *The interfaces file with eth0 setup for a static IP address*

Notice the changes that were made to the eth0 section. Instead of deleting the old iface line, you can just use a # symbol to comment it out so that it is easier to reverse your changes if something goes wrong. The type of configuration has been changed from manual to static. Added under the iface line are the required parameters for a static IP address: an address, a network mask, a network gateway address, and DNS server addresses.

WiFi

In order to use WiFi, we just need to make a few changes to the wlan0 section (see Figure 4-17).

```
# interfaces(5) file used by ifup(8) and ifdown(8)

# Please note that this file is written to be used with dhcpcd
# For static IP, consult /etc/dhcpcd.conf and 'man dhcpcd.conf'

# Include files from /etc/network/interfaces.d:
source-directory /etc/network/interfaces.d

auto lo
iface lo inet loopback

#iface eth0 inet manual
iface eth0 inet static
    address 192.168.0.100
    netmask 255.255.255.0
    gateway 192.168.0.1
    dns-nameservers 8.8.8.8 8.8.4.4

allow-hotplug wlan0
iface wlan0 inet manual
#    wpa-conf /etc/wpa_supplicant/wpa_supplicant.conf
    wpa-ssid "YourSSID"
    wpa-psk "YourPassword"

allow-hotplug wlan1
iface wlan1 inet manual
    wpa-conf /etc/wpa_supplicant/wpa_supplicant.conf
```

Figure 4-17. *The interfaces file with wlan0 setup for simple WiFi access*

Here, I've commented out the line for wpa-conf and added a line to configure the SSID of my wireless access point as well as the password. This is by far the simplest way to get quick access to WiFi. If you wanted a static IP on this interface, you could make the same changes that we made for the eth0 interface in addition to the SSID and password configuration.

More Secure WiFi with Multiple Networks

Sometimes you need more control over your WiFi interfaces. For example, you might want to automatically switch between home and work access points without changing the configuration each time. You might have a hidden SSID that you need to connect to or an enterprise password encryption scheme you need to use. When things get more complex, it's best to use the *wpa_supplicant.conf* file in addition to the *interfaces* file.

WPA stands for *WiFi Protected Access*. WPA adds more security protocols to make WiFi more secure and harder to break into. Let's start by reverting our *interfaces* file to the default to make use of the *wpa_supplicant.conf* file (see Figure 4-18).

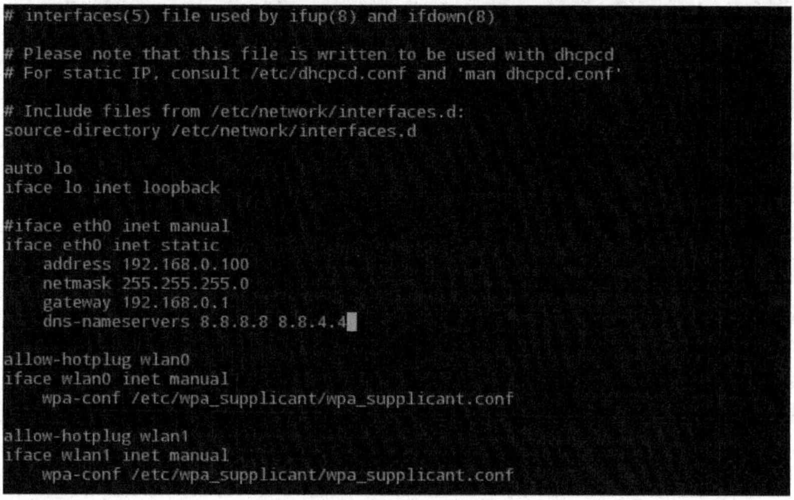

```
# interfaces(5) file used by ifup(8) and ifdown(8)

# Please note that this file is written to be used with dhcpcd
# For static IP, consult /etc/dhcpcd.conf and 'man dhcpcd.conf'

# Include files from /etc/network/interfaces.d:
source-directory /etc/network/interfaces.d

auto lo
iface lo inet loopback

#iface eth0 inet manual
iface eth0 inet static
    address 192.168.0.100
    netmask 255.255.255.0
    gateway 192.168.0.1
    dns-nameservers 8.8.8.8 8.8.4.4

allow-hotplug wlan0
iface wlan0 inet manual
    wpa-conf /etc/wpa_supplicant/wpa_supplicant.conf

allow-hotplug wlan1
iface wlan1 inet manual
    wpa-conf /etc/wpa_supplicant/wpa_supplicant.conf
```

Figure 4-18. *The interfaces file with wlan0 changes removed*

Uncommenting the `wpa-conf` line and deleting the `wpa-ssid` and `wpa-psk` lines means the wlan0 interface will now refer to the *wpa_supplicant.conf* file for its configuration. Figure 4-19 shows what that file looks like by default on the Raspberry Pi.

```
ctrl_interface=DIR=/var/run/wpa_supplicant GROUP=netdev
update_config=1
country=GB
```

Figure 4-19. *The default wpa_supplicant.conf file*

The `country` setting should be changed automatically if you set up your internationalization settings in Chapter 1. If not, you can find your country code by searching the ISO website (*https://www.iso.org/obp/ui/search*).

Updating this file with one or more network sections will allow your WiFi interface to connect to the network. In Figure 4-20, I have updated the file with the necessary information for most situations.

```
ctrl_interface=DIR=/var/run/wpa_supplicant GROUP=netdev
update_config=1
country=GB

network={
    ssid="YourSSID"
    scan_ssid=0
    psk="YourPassword"
    key_mgmt=WPA-PSK
}
```

Figure 4-20. *The wpa_supplicant file with an added network section*

The `ssid` and `psk` options represent the SSID name and password. The `scan_ssid=0` line lets the systems know that this is not a hidden network. If it were a hidden network, you would need to change the value from 0 to 1. The `key_mgmt=WPA-PSK` line represents the password encryption your access point is using. WPA-PSK should work for most home users. If you are working in an office environment, you may need to change this to something else. You can find out about all the encryption types that `wpa_supplicant` supports by referencing the manpage for *wpa_supplicant.conf.*

You can add further network sections to be able to connect to different networks depending on your location (see Figure 4-21). `wpa_supplicant` will automatically detect the best network based on availability and signal strength.

```
ctrl_interface=DIR=/var/run/wpa_supplicant GROUP=netdev
update_config=1
country=GB

network={
    ssid="YourSSID"
    scan_ssid=0
    psk="YourPassword"
    key_mgmt=WPA-PSK
}
network={
    ssid="OtherSSID"
    scan_ssid=0
    psk="OtherPassword"
    key_mgmt=WPA-PSK
}
```

Figure 4-21. *The wpa_supplicant file with two network sections*

When you're done configuring your network, the easiest way to implement your changes is simply to reboot your Raspberry Pi.

Installing Software: apt

Adding software to Linux is different from other operating systems. Since much of the software you run on Linux is open source and free, public repositories of software packages are maintained for the various distributions of Linux that exist. A software package manager is used to download a package, install or remove a package, manage any dependencies on other software that may exist, and keep packages up-to-date.

Open Source Software

Open source software is different than proprietary or closed sourced software in many ways. First and foremost, as the word *open* implies, the source code for the software is available for anyone to look at and inspect. Second, because the code is available, this naturally invites contributions from the community. If there's a problem with the software, you can correct it yourself by submitting a bug report or a patch that fixes the issue. Third, you can share and distribute open source software to others without breaking the law or violating some sort of license agreement. In fact, this sharing behavior is generally encouraged. This is in stark contrast to what you may typically think of when it comes to sharing content (see Appendix A for more information on the history of open source software).

Because the Rasbian distribution used most often on the Raspberry Pi is based on Debian Linux, it uses the Debian package manager software called apt, which stands for *Advanced Package Tool*. apt contains a set of tools that can be used to perform various tasks related to software package management. The most frequently used tool is called apt-get, and it handles almost all the functionality you will need when it comes to installing software with the exception of searching, which is done with apt-cache.

Since apt can significantly alter your system, you are required to run some of the apt tools using sudo.

Using apt-get update

There are thousands of software packages for Linux and they are updated frequently. In fact, if you checked on the software updates for your Raspberry Pi, you would find that there are several updates a day. Now, that's not to say that you need to update your software every day. Most updates are enhancements or minor bug fixes, and not having them won't mess anything up for you. But sometimes there are updates that are related to system security, and those can be important. A good rule of thumb is to check for software updates once a month or so. You also want to check for updates right before you install any software to make sure the software database on your Raspberry Pi is current.

To check for software updates, type the following on the console:

```
sudo apt-get update
```

This will download the list of software from the repositories that have been preconfigured in the system. The list will then be read and the software database will be updated to include information about new and updated software packages (see Figure 4-22).

```
pi@raspberrypi ~ $ sudo apt-get update
Get:1 http://archive.raspberrypi.org jessie InRelease [13.2 kB]
Get:2 http://mirrordirector.raspbian.org jessie InRelease [14.9 kB]
Get:3 http://archive.raspberrypi.org jessie/main Sources [50.3 kB]
Get:4 http://mirrordirector.raspbian.org jessie/main armhf Packages [8,982 kB]
Get:5 http://archive.raspberrypi.org jessie/ui Sources [10.6 kB]
Get:6 http://archive.raspberrypi.org jessie/main armhf Packages [148 kB]
Get:7 http://archive.raspberrypi.org jessie/ui armhf Packages [14.8 kB]
Ign http://archive.raspberrypi.org jessie/main Translation-en_US
Ign http://archive.raspberrypi.org jessie/main Translation-en
Ign http://archive.raspberrypi.org jessie/ui Translation-en_US
Ign http://archive.raspberrypi.org jessie/ui Translation-en
Get:8 http://mirrordirector.raspbian.org jessie/contrib armhf Packages [37.5 kB]
Get:9 http://mirrordirector.raspbian.org jessie/non-free armhf Packages [70.3 kB
]
Get:10 http://mirrordirector.raspbian.org jessie/rpi armhf Packages [1,356 B]
Ign http://mirrordirector.raspbian.org jessie/contrib Translation-en_US
Ign http://mirrordirector.raspbian.org jessie/contrib Translation-en
Ign http://mirrordirector.raspbian.org jessie/main Translation-en_US
Ign http://mirrordirector.raspbian.org jessie/main Translation-en
Ign http://mirrordirector.raspbian.org jessie/non-free Translation-en_US
Ign http://mirrordirector.raspbian.org jessie/non-free Translation-en
Ign http://mirrordirector.raspbian.org jessie/rpi Translation-en_US
Ign http://mirrordirector.raspbian.org jessie/rpi Translation-en
Fetched 9,343 kB in 28s (332 kB/s)
Reading package lists... Done
pi@raspberrypi ~ $
```

Figure 4-22. *Typical output of the apt-get update command*

Using apt-get upgrade

Once you've updated your software database, you're ready to either upgrade the software you have installed or install new software. *Upgrade* actually installs new versions of your software, whereas *update* only updates the software database. Upgrade your software by typing the following command:

```
sudo apt-get upgrade
```

The first thing that happens during an upgrade is that apt-get will read the package list and check for dependencies. Because software in Linux is open source, it can be built in a modular fashion. If someone else has already built a program that does a certain function, other programs can simply use it instead of rebuilding the functionality from scratch every time. In this way, one program becomes dependent on another program to function correctly, and it is important to manage these dependencies so that everything works properly.

After checking for dependencies, apt-get will calculate which packages need to be updated and output a list of those packages. It will also show the total size of the download and the amount of total disk space the updates will occupy after they're installed (see Figure 4-23).

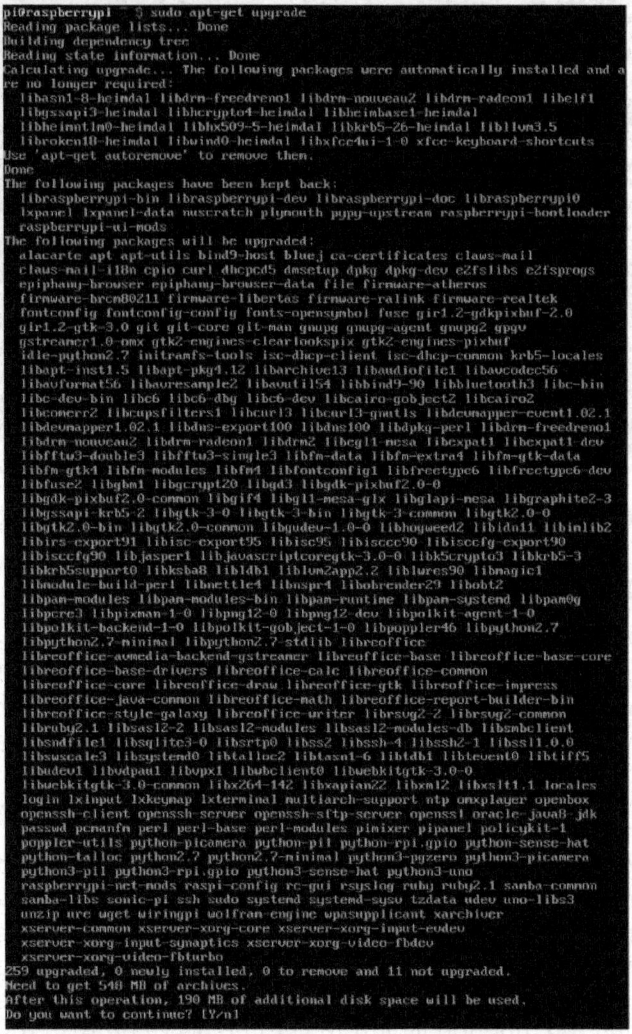

Figure 4-23. *Example of a rather large upgrade using apt-get upgrade*

As you can see, it's been a while since I have upgraded the software on this particular Raspberry Pi. If you have any concerns about the update, you can press N to exit. Otherwise, type **Y** or just press Enter to continue.

Be aware that when you continue from this point, `apt-get` will begin downloading the software packages and then start to install them. It could take quite a long time (we're talking many minutes to a few hours) if you have a lot of updates, are using a slower Raspberry Pi like the Raspberry Pi 1 or Zero, or have a slow network connection (see Figure 4-24).

```
Get:203 http://archive.raspberrypi.org/debian/ jessie/main libegl1-mesa armhf 11
.1.0-1+rpi1 [90.1 kB]
Get:204 http://archive.raspberrypi.org/debian/ jessie/main libfm-extra4 armhf 1.
2.3-1+rpi1 [28.4 kB]
Get:205 http://archive.raspberrypi.org/debian/ jessie/main libfm-data all 1.2.3-
1+rpi1 [216 kB]
Get:206 http://archive.raspberrypi.org/debian/ jessie/main libfm-gtk4 armhf 1.2.
3-1+rpi1 [149 kB]
Get:207 http://archive.raspberrypi.org/debian/ jessie/main libfm4 armhf 1.2.3-1+
rpi1 [105 kB]
Get:208 http://archive.raspberrypi.org/debian/ jessie/main libfm-gtk-data all 1.
2.3-1+rpi1 [32.0 kB]
Get:209 http://archive.raspberrypi.org/debian/ jessie/main libgtk-3-common all 3
.14.5-1+deb8u1rpi1rpi1g [3,061 kB]
Get:210 http://archive.raspberrypi.org/debian/ jessie/main libgtk-3-0 armhf 3.14
.5-1+deb8u1rpi1rpi1g [1,884 kB]
Get:211 http://archive.raspberrypi.org/debian/ jessie/main libwebkitgtk-3.0-comm
on all 1:2.4.1-1rpi53rpi1g [462 kB]
Get:212 http://archive.raspberrypi.org/debian/ jessie/main libwebkitgtk-3.0-ar
mhf 1:2.4.1-1rpi53rpi1g [5,212 kB]
Get:213 http://archive.raspberrypi.org/debian/ jessie/main libjavascriptcoregtk-
3.0-0 armhf 1:2.4.1-1rpi53rpi1g [1,716 kB]
Get:214 http://archive.raspberrypi.org/debian/ jessie/main libvdpau1 armhf 1.1.1
-1~bpo8+1 [38.6 kB]
Get:215 http://archive.raspberrypi.org/debian/ jessie/main oracle-java8-jdk armh
f 8u65 [62.2 MB]
Get:216 http://archive.raspberrypi.org/debian/ jessie/main wolfram-engine armhf
10.3.1+2016012407 [236 MB]
55% [216 wolfram-engine 56.5 MB/236 MB 24%]                     1,348 kB/s 3min 3s
```

Figure 4-24. *Example of a rather large upgrade using apt-get upgrade (continued)*

If your console screen isn't wide enough, the text will wrap around as the packages are downloaded. The percentage on the far left of the screen shows the overall download progress. Once all the packages are downloaded, `apt-get` will begin unpacking, processing, and setting them up one by one. When the process is complete, you will be returned to the prompt (see Figure 4-25).

```
Setting up rc-gui (1.1-1) ...
Setting up libruby2.1:armhf (2.1.5-2+deb8u3) ...
Setting up ruby2.1 (2.1.5-2+deb8u3) ...
Setting up ruby (1:2.1.5+deb8u2) ...
Setting up sudo (1.8.10p3-1+deb8u3) ...
Setting up unzip (6.0-16+deb8u2) ...
Setting up wiringpi (2.32) ...
Setting up wpasupplicant (2.3-1+deb8u4) ...
Setting up xarchiver (1:0.5.4-1+deb8u1) ...
Setting up xserver-common (2:1.17.2-1+rpi1) ...
Setting up xserver-xorg-core (2:1.17.2-1+rpi1) ...
Setting up xserver-xorg-video-fbturbo (1.20150305~205709) ...
Setting up xserver-xorg-video-fbdev (1:0.4.4-1+rpi1) ...
Setting up xserver-xorg-input-evdev (1:2.9.2-1~bpo8+1) ...
Setting up xserver-xorg-input-synaptics (1.8.2-1~bpo8+1) ...
Setting up bluej (3.1.7) ...
Setting up python-picamera (1.12) ...
Setting up python3-picamera (1.12) ...
Setting up raspberrypi-net-mods (1.2.3) ...
Updating /etc/network/interfaces. Original backed up as interfaces.dpkg-old.
Setting up sonic-pi (1:2.10.0-2) ...
Enabling /etc/security/limits.d/audio.conf for jackd2
Processing triggers for initramfs-tools (0.120+deb8u2) ...
Processing triggers for libc-bin (2.19-18+deb8u6) ...
Processing triggers for libgdk-pixbuf2.0-0:armhf (2.31.1-2+deb8u5) ...
Processing triggers for ca-certificates (20141019+deb8u1) ...
Updating certificates in /etc/ssl/certs... 19 added, 18 removed; done.
Running hooks in /etc/ca-certificates/update.d....done.
Processing triggers for systemd (215-17+deb8u5) ...
pi@raspberrypi ~ $
```

Figure 4-25. *Example of a rather large upgrade using apt-get upgrade (continued)*

This particular upgrade took about 40 minutes to complete on an older Raspberry Pi with a slower-than-average network connection. More frequent upgrades on the order of once a month will prevent the upgrade process from taking so long. Although not technically required, it's a good idea to reboot your system after upgrading it, especially when the upgrade is a large one.

Using apt-cache

With so many software packages available for Linux, it can be difficult to remember their names. There is a way to search through the database to find the package you're looking for. The tool apt-cache can be used to search through the software package database and even show useful information about individual packages. To search for a package, type:

```
apt-cache search pattern
```

apt-cache will search through the database and return any package name and description that contains the pattern you supplied (see Figure 4-26).

```
pi@raspberrypi ~ $ apt-cache search gpio
ledmon - Enclosure LED Utilities
stm32flash - STM32 chip flashing utility using a serial bootloader
pigpio - Library for Raspberry Pi GPIO control
python-gpiozero - Simple API for controlling devices attached to the GPIO pins.
python-gpiozero-doc - Documentation for the gpiozero API
python-pigpio - Python module which talks to the pigpio daemon (Python 2)
python-rpi.gpio - Python GPIO module for Raspberry Pi
python-w1thermsensor - Python w1 therm sensor module (Python 2)
python3-gpiozero - Simple API for controlling devices attached to the GPIO pins.
python3-pigpio - Python module which talks to the pigpio daemon (Python 3)
python3-rpi.gpio - Python 3 GPIO module for Raspberry Pi
python3-w1thermsensor - Python w1 therm sensor module (Python 3)
raspi-gpio - Dump the state of the BCM270x GPIOs
wiringpi - The wiringPi libraries, headers and gpio command
pi@raspberrypi ~ $ _
```

Figure 4-26. *Example output of the apt-cache search command*

Once you find the package name you're looking for, it can be helpful to get more information about it. You can use apt-cache to display more detailed information by using the show function:

```
apt-cache show pigpio
```

You can see who wrote the software, the current version, how much disk space it takes up, what the website for the software package is, a detailed description, and more (see Figure 4-27).

```
pi@raspberrypi ~ $ apt-cache show pigpio
Package: pigpio
Version: 1.30-1
Architecture: armhf
Maintainer: Serge Schneider <serge@raspberrypi.org>
Installed-Size: 837
Depends: libc6 (>= 2.17), init-system-helpers (>= 1.18~)
Homepage: http://abyz.co.uk/rpi/pigpio/
Priority: optional
Section: utils
Filename: pool/main/p/pigpio/pigpio_1.30-1_armhf.deb
Size: 229000
SHA256: 6c61526aeb0389fd7b021cd1cce54562a21408699d0b179057134a17e4f832e0
SHA1: fff1c4b463b1f97d00e19e003d0797b3fb5754d8
MD5sum: 6134ceaeb2f3cb053bf59f6fa637beb8
Description: Library for Raspberry Pi GPIO control
 Library for the Raspberry which allows control of the General Purpose Input Ou
tputs (GPIO).
 .
 pigpio is written in C but may be used by other languages.
 In particular the pigpio daemon offers a socket and pipe interface to the unde
rlying library.
Description-md5: 95c13711224672ba0acd34af4b9ac647

pi@raspberrypi ~ $ _
```

Figure 4-27. *Example output of the apt-cache show command*

Using apt-get install

Installing new software is handled with the `apt-get install` command. You can install multiple packages at one time and `apt-get` will manage installing any required dependencies for you. To install software on your Raspberry Pi, just type:

> sudo apt-get update
>
> sudo apt-get install *packagename packagename packagename*

`apt-get` will let you know how much data will be downloaded and how much disk space will be used after the install is complete. If there is only one package to be installed, `apt-get` will not ask you to continue and will install the package without prompting (see Figure 4-28).

 Skip the Confirmation Message

You can avoid being prompted to confirm whether you want to install the software packages by using the -y option flag (i.e., `sudo apt-get -y install` *pack agename packagename packagename*).

```
pi@raspberrypi   $ sudo apt-get install pigpio
Reading package lists... Done
Building dependency tree
Reading state information... Done
The following NEW packages will be installed:
  pigpio
0 upgraded, 1 newly installed, 0 to remove and 11 not upgraded.
Need to get 0 B/229 kB of archives.
After this operation, 857 kB of additional disk space will be used.
Selecting previously unselected package pigpio.
(Reading database ... 110763 files and directories currently installed.)
Preparing to unpack .../pigpio_1.30-1_armhf.deb ...
Unpacking pigpio (1.30-1) ...
Processing triggers for man-db (2.7.0.2-5) ...
Setting up pigpio (1.30-1) ...
pi@raspberrypi   $ _
```

Figure 4-28. *Example output of the apt-get install command*

Notice how man was updated with a new page about this program on the next-to-last line.

apt-get remove

Similar to installing software, removing packages is a straightforward process. To remove packages, simply type:

    sudo apt-get remove *packagename packagename packagename*

apt-get will remove the packages and the associated manpage entries (see Figure 4-29).

```
pi@raspberrypi   $ sudo apt-get remove pigpio
Reading package lists... Done
Building dependency tree
Reading state information... Done
The following packages will be REMOVED:
  pigpio
0 upgraded, 0 newly installed, 1 to remove and 11 not upgraded.
After this operation, 857 kB disk space will be freed.
Do you want to continue? [Y/n]
(Reading database ... 110787 files and directories currently installed.)
Removing pigpio (1.30-1) ...
Processing triggers for man-db (2.7.0.2-5) ...
pi@raspberrypi   $ _
```

Figure 4-29. *Example output of the apt-get remove command*

Linux will not remove all the dependencies, however. You will notice that after you install or upgrade software, apt-get might mention that there are packages that are no longer required. You can uninstall these by using the following command:

    sudo apt-get autoremove

apt-get will show you how much disk space will be freed up and prompt you to confirm. It will then remove all the programs that are no longer required by any other software packages (see Figure 4-30).

```
pi@raspberrypi ~ $ sudo apt-get autoremove
Reading package lists... Done
Building dependency tree
Reading state information... Done
The following packages will be REMOVED:
  libasn1-8-heimdal libdrm-freedreno1 libdrm-nouveau2 libdrm-radeon1 libelf1
  libgssapi3-heimdal libhcrypto4-heimdal libheimbase1-heimdal
  pigpio
0 upgraded, 0 newly installed, 1 to remove and 11 not upgraded.
After this operation, 857 kB disk space will be freed.
Do you want to continue? [Y/n]
(Reading database ... 110787 files and directories currently installed.)
Removing pigpio (1.30-1) ...
Processing triggers for man-db (2.7.0.2-5) ...
pi@raspberrypi ~ $
```

Figure 4-30. *Example output of the apt-get autoremove command*

apt-get dist-upgrade

There is a special type of upgrade you can perform that will upgrade your whole distribution to the latest and greatest version. You perform this with the command apt-get dist-upgrade. Like an install or upgrade operation, you need to update the software database with apt-get update before you run this. The process is very similar to doing a regular upgrade, but could take quite a bit longer since more packages will need to be downloaded and installed. The benefit of upgrading to the latest distribution is that it would allow you to take advantage of new features that aren't available in your version. All your personal files and configuration should remain intact.

However, don't feel like you *have to* upgrade your distribution just because a new version is available. If everything is working fine, be content to stay where you are. Updating your distribution is not like updating the software version on your smartphone. Very rarely is there a must-have feature that you really need to make your project work. Your version of Linux should work fine and be supported with security updates for at least a few years. To make sure you have those security updates, you can simply use `apt-get update` instead of `apt-get dist-upgrade`.

One thing to note is that upgrading your distribution this way may not install software packages that aren't strictly required. In some cases, it might be easier to start from scratch with the latest released image file as we did in Chapter 1. Just be sure to back up your files first.

Fixing Conflicts

Occasionally, you will get an error that mentions "missing dependencies" or "broken packages" when `apt-get` is trying to install software. This usually means that you haven't updated in a while. To fix this, you should first try running:

```
sudo apt-get update
sudo apt-get upgrade
```

This will update your software repository and upgrade your installed packages to make sure everything is up-to-date. You can then try to install your software again.

Try It for Yourself

Install the pigpio main library and the associated python pigpio library. In order to do this, you will need to find out the names of the software packages and confirm they are the right ones by reading the description. The first step is to update the package database and then look up the names of the packages:

```
sudo apt-get update

apt-cache search pigpio
```

You should get a search result that shows all the packages with *pigpio* in the title or description (see Figure 4-31).

```
pi@raspberrypi ~ $ apt-cache search pigpio
pigpio - Library for Raspberry Pi GPIO control
python-pigpio - Python module which talks to the pigpio daemon (Python 2)
python3-pigpio - Python module which talks to the pigpio daemon (Python 3)
pi@raspberrypi ~ $
```

Figure 4-31. *Using apt-cache search to find a software package*

Next, install both packages at the same time without being prompted. Let's assume we will be using Python 2 to write some programs and not Python 3 for now. Use the `apt-get` command to install the software (see Figure 4-32):

```
sudo apt-get -y pigpio python-pigpio
```

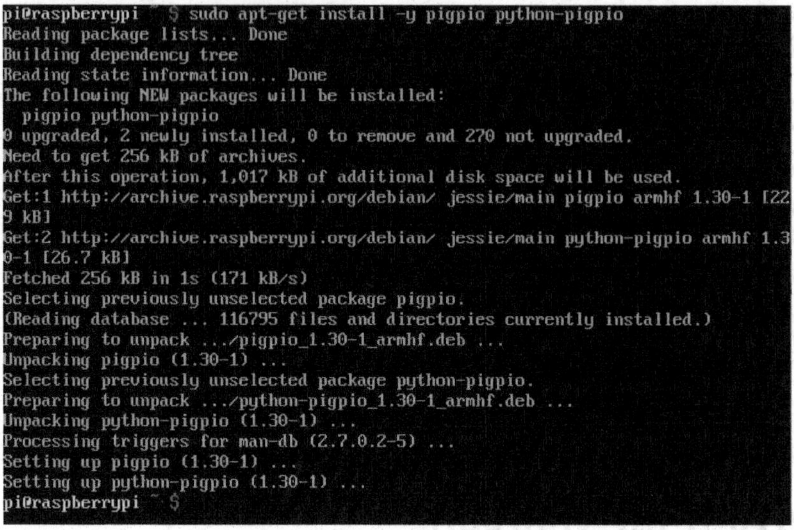

```
pi@raspberrypi ~ $ sudo apt-get install -y pigpio python-pigpio
Reading package lists... Done
Building dependency tree
Reading state information... Done
The following NEW packages will be installed:
  pigpio python-pigpio
0 upgraded, 2 newly installed, 0 to remove and 270 not upgraded.
Need to get 256 kB of archives.
After this operation, 1,017 kB of additional disk space will be used.
Get:1 http://archive.raspberrypi.org/debian/ jessie/main pigpio armhf 1.30-1 [22
9 kB]
Get:2 http://archive.raspberrypi.org/debian/ jessie/main python-pigpio armhf 1.3
0-1 [26.7 kB]
Fetched 256 kB in 1s (171 kB/s)
Selecting previously unselected package pigpio.
(Reading database ... 116795 files and directories currently installed.)
Preparing to unpack .../pigpio_1.30-1_armhf.deb ...
Unpacking pigpio (1.30-1) ...
Selecting previously unselected package python-pigpio.
Preparing to unpack .../python-pigpio_1.30-1_armhf.deb ...
Unpacking python-pigpio (1.30-1) ...
Processing triggers for man-db (2.7.0.2-5) ...
Setting up pigpio (1.30-1) ...
Setting up python-pigpio (1.30-1) ...
pi@raspberrypi ~ $
```

Figure 4-32. *Installing pigpio with apt-get*

Rebooting and Shutting Down

The first time I ran Linux without a desktop, I ran into a problem. Everything was going fine until I needed to shut down the system. I was still very new to Linux at the time and I didn't know how to start the desktop up again (see Chapter 5 for how to do this). Eventually, I ended up shutting off the power to the system. This is a very bad idea on any computer, but especially bad on the Raspberry Pi.

As I mentioned in Chapter 2, everything in Linux is represented by a file, including the state of the operating system. Linux is constantly writing to these files as updates occur on the system. If you kill power to your Raspberry Pi while the system is in the middle of updating a file, you will end up with a corrupt file that can't be read when the system is powered on again. Depending on what file is corrupted, you might lose some of the project files you were working on, or worse, you might end up being unable to boot up your Raspberry Pi again. This problem is compounded by the fact that the Raspberry Pi uses an SD card for storage and, compared to other storage devices, the write speed of SD cards is still pretty slow.

So it is important that you know how to reboot and shut down your system properly from the command line. The command you use to do either of these is **shutdown**. The **shutdown** command should work on almost all Linux and Unix systems. To reboot a Linux system from the command line, type:

```
sudo shutdown -r now
```

To shut down a Linux system, type:

```
sudo shutdown -h now
```

Notice that the only difference between the two commands is using **-r** to reboot or **-h** to halt the system and shut it down entirely. If you are shutting down your system, you will know when the shutdown is complete when the LED on the board blinks on and off 10 times. After that, it is safe to unplug your Raspberry Pi.

Why This Matters for Makers

Knowing the basics of using the command line make it easier to navigate around in Linux, get connected to the internet, and install software. These operations are the bare minimum that a Maker should know before venturing out on their own beyond an online tutorial to start building their own really cool projects. You can go further and impress your family and friends by becoming a command-line wizard with the tips and tricks I will show you in Chapter 6.

5/Headless Operation

In this chapter I will explain how to connect to a Raspberry Pi running Linux over a network without a keyboard, mouse, or monitor attached to it (aka *headless*). The ability to operate and interact with a project remotely is important for any Maker, and opens up new possibilities that would not be available if you always needed a keyboard, mouse, and monitor to be directly attached to an SBC running Linux. This is perfect for projects that you want to be mobile or that are just meant to run quietly in the corner collecting or serving data.

Turning Off the Desktop

Most of the time, a Raspberry Pi running headless doesn't need the desktop running. Turning off the desktop is a relatively easy process and is configured through the `raspi-config` utility. On the console, run this command:

```
sudo raspi-config
```

This will open up the Raspberry Pi configuration application, as shown in Figure 5-1. Use the arrow keys to move the selection cursor down to Boot Options and press Enter.

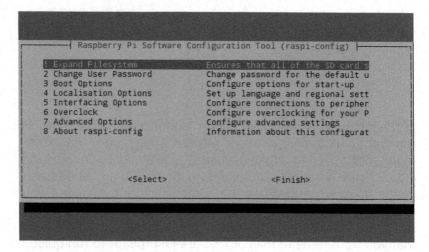

Figure 5-1. *The raspi-config main menu*

Choose Desktop/CLI and press Enter (see Figure 5-2).

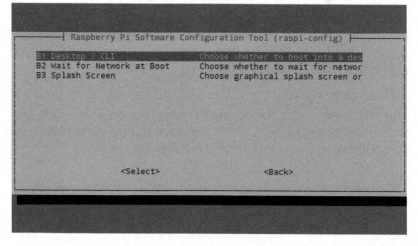

Figure 5-2. *The raspi-config Boot Options menu*

Now choose Console if you want to be forced to log in when the system boots up, or Console Autologin, which will automatically log in the user "pi" for you. Pressing Enter on your selection will return you to the main `raspi-config` menu (see Figure 5-3).

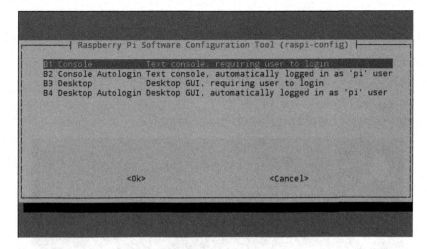

Figure 5-3. *The raspi-config Desktop/CLI menu*

Press Tab to move the selection cursor to Finish and press the Enter key to exit. You will be asked if you want to reboot. Whether you reboot now or later, the next time you do, you will be brought right to the console instead of the desktop.

If you ever want to get back into the desktop for any reason, you can run the following on the command line:

```
startx
```

If you ever want to change back to automatically boot into the desktop, just go through these steps again and choose one of the desktop options in the `raspi-config` boot options menu.

Finding Your System on the Network

In order to connect to your Raspberry Pi, you need to know its *IP address*. An IP address is a unique identifier assigned to every computer on your network. You can find your IP address on the Pi itself, from the router on your network, or from an app on your phone.

Raspberry Pi

The easiest way to find the IP address of your Raspberry Pi is by looking it up before you disconnect your monitor, keyboard, and mouse. This can be done from the command line or desktop.

When the Raspberry Pi boots up, it should show you the IP address just before you get to the prompt (see Figure 5-4).

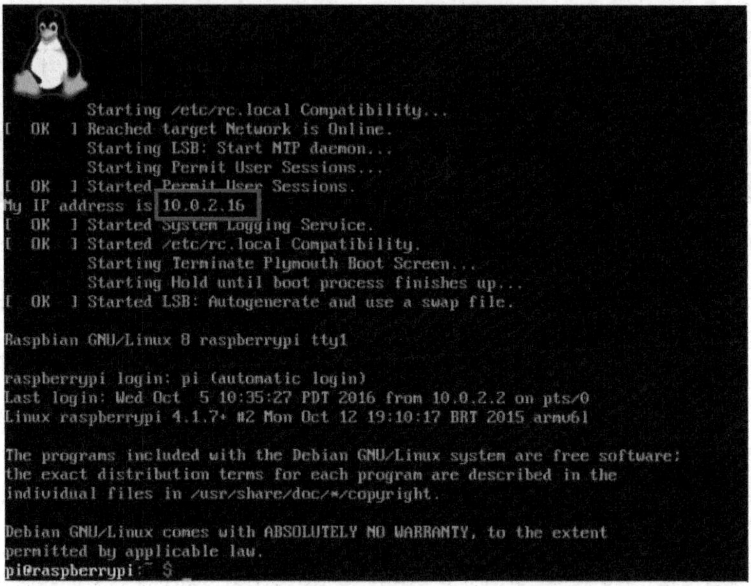

Figure 5-4. *Finding the Raspberry Pi IP address when it boots up*

If you've been using your Raspberry Pi and can't see this information anymore, you can find the IP address by running the command `ip addr show` (see Figure 5-5).

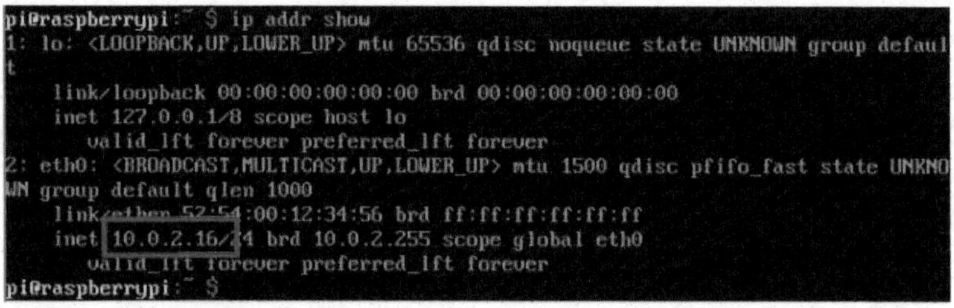

Figure 5-5. *Finding the Raspberry Pi IP address with ip addr show*

Use the address listed under the eth0 section if your system is connected to the network via a wired Ethernet cable. If your system is connected via WiFi, use the address listed under wlan0. In the preceding example, we would use 10.0.2.16 without any trailing slashes or other numbers.

Router

Most modern routers will show you the connected devices on your network either in list form or as a network map. Connect to your router's built-in website and find the configuration page to view this information. In my case, the router shows a network map (see Figure 5-6).

Figure 5-6. *Network map from a home router*

By clicking on the device labeled raspberrypi, I can see its IP address (see Figure 5-7). In this case, it's 192.168.0.209.

Figure 5-7. *Device details from home router*

Android/iPhone

There are many apps for Android and iPhone that can scan your network and find devices and IP addresses. The one I currently use is called "Fing - Network Tools." You can find it in the Google Play Store and the Apple App Store. Once the app is installed on your phone, simply open it and click on the refresh button to find your Raspberry Pi (see Figure 5-8).

Figure 5-8. *Fing app showing Raspberry Pi IP address*

Command-Line Access: ssh

Just because your project is running without a keyboard and mouse doesn't mean that you don't need access to it. You will need to upload files, change the configuration, and most importantly, be able to shut down the system gracefully from time to time. The best way to access the command line remotely is with a tool called SSH.

SSH stands for *secure shell*. As you can guess by the name, it provides secure access to the shell on a remote system. SSH is different than its predecessor telnet, which was the standard for many years. Unfortunately, telnet traffic was sent completely "in the clear"—usernames and passwords were easily readable by any computer the data passed through. SSH is secure because the communication to the remote system is encrypted so that it can't be read by other systems on the network.

For SSH to work, there needs to be two components: an SSH client on your local computer and an SSH server on the remote system. The Raspberry Pi already has an SSH server installed on it as part of the Raspbian distribution of Linux. However, for security, the server is not running by default. You can turn it on by using the `raspi-config` tool (see Figure 5-9):

```
sudo raspi-config
```

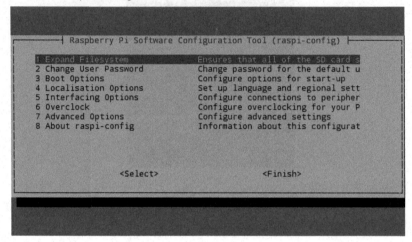

Figure 5-9. *The raspi-config main menu*

Use the arrow keys to move the red selection cursor down to Interfacing Options and press Enter. In the Interfacing Options menu, use the arrow keys to move the red selection cursor down to SSH and press Enter (see Figure 5-10).

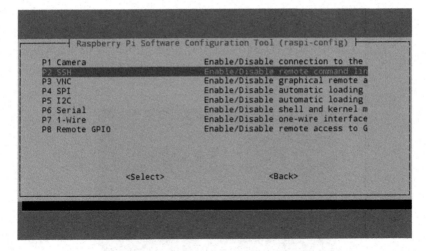

Figure 5-10. *The raspi-config Interfacing Options menu*

The next screen will ask you if you would like to enable the SSH server. Choose Yes. Once the SSH server is enabled, select OK to return to the main `raspi-config` tool screen. Press the Tab key to move the selection cursor to Finish and press the Enter key to exit.

Now you need to install an SSH client on your computer so that you can connect to the SSH server on the Raspberry Pi. In the following sections, I will recommend how to connect to a Raspberry Pi from Windows, macOS, Linux, and Android. Although I won't cover software installation, I will give examples of how to connect and what to expect on each platform.

Windows

Windows does not have a built-in SSH client, so you need to install one. One of the most widely used SSH clients is PuTTY. You can download the latest version of PuTTY online (*http://bit.ly/ukputty*). Once you've installed PuTTY, you can open it from the Windows start menu (see Figure 5-11).

Which PuTTY Do I Use?

The default installation of PuTTY comes with several programs. The one you want is simply called PuTTY.

Figure 5-11. *Launching the PuTTY application from Windows*

What Is TTY?

TTY comes from the word *TeleTYpe* and refers to a way of communicating with a computer. Back in the early days of computing, you needed to use a teletypewriter or teleprinter machine to type out the information that you wanted to feed into the computer. You can still see the remnants of this technology in Linux and in other remote communications programs like PuTTY.

Once you've opened PuTTY, type the IP address of your Raspberry Pi in the box labeled "Host Name (or IP address)" (see Figure 5-12). Click the Open button to launch your SSH session.

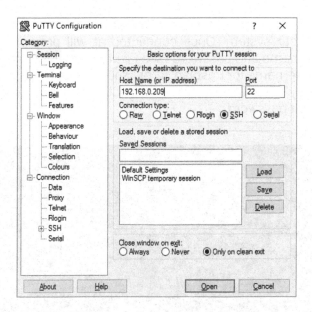

Figure 5-12. *The PuTTY configuration screen*

The first time you connect to your remote system, you will be asked to verify the SSH server's encryption key (see Figure 5-13). Click Yes.

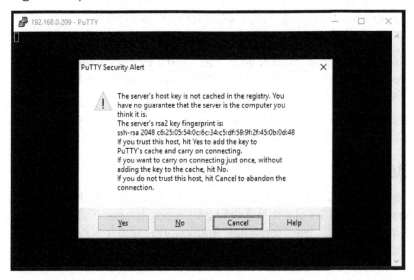

Figure 5-13. *PuTTY security warning*

PuTTY will then open a black terminal window and connect to your remote system. Once connected, it will prompt you for your username and password (see Figure 5-14). Remember: the default username for Rasbian is "pi," and the default password is "raspberry." See Chapter 2 for instructions on how to change the default password.

Figure 5-14. *The login prompt to connect with PuTTY*

If you get an error screen that says "Network error: Connection timed out," that means your Raspberry Pi is not reachable on your network. Close the console window, check your IP address, check the connections to your Pi, and try again.

Once you're connected, you'll be at the prompt and ready to start entering commands.

MacOS

Because macOS is based on UNIX, it makes sense that there is already an SSH client installed and ready to use. To get to it, in Finder, open the Terminal application under Utilities (see Figure 5-15).

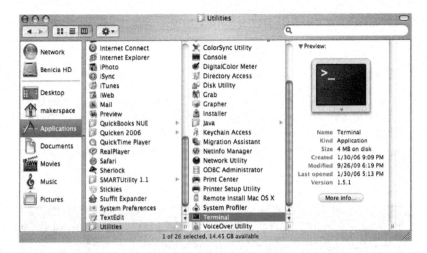

Figure 5-15. *Using Finder to locate the Terminal application*

Once open, simply type **ssh *user@ip-address*** to connect to your Raspberry Pi (see Figure 5-16).

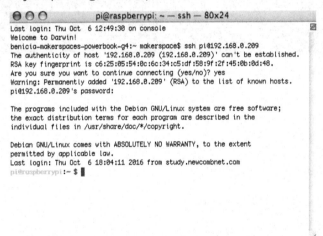

Figure 5-16. *Using ssh to connect to a Raspberry Pi from macOS*

If this is the first time you've connected to this remote system, you'll be prompted to accept the SSH server's encryption key. Type **yes** and press the Enter key to continue. Once you're connected, you will be at the prompt and ready to start entering commands.

Linux

There is already an SSH client installed in Linux. All you need to do is type **ssh** *user@ip-address* from the console or a terminal emulator (see Figure 5-17). Since the Raspberry Pi is running Linux, you can even use a Raspberry Pi to ssh into another Raspberry Pi!

    ssh pi@192.168.0.209

```
pi@raspberrypi:~ $ ssh pi@192.168.0.209
The authenticity of host '192.168.0.209 (192.168.0.209)' can't be established.
ECDSA key fingerprint is 6e:d6:5e:05:93:47:a3:3e:1d:f2:ee:66:76:75:29:35.
Are you sure you want to continue connecting (yes/no)? yes
Warning: Permanently added '192.168.0.209' (ECDSA) to the list of known hosts.
pi@192.168.0.209's password:

The programs included with the Debian GNU/Linux system are free software;
the exact distribution terms for each program are described in the
individual files in /usr/share/doc/*/copyright.

Debian GNU/Linux comes with ABSOLUTELY NO WARRANTY, to the extent
permitted by applicable law.
Last login: Thu Oct  6 17:58:01 2016 from study.newcombnet.com
pi@raspberrypi:~ $
```

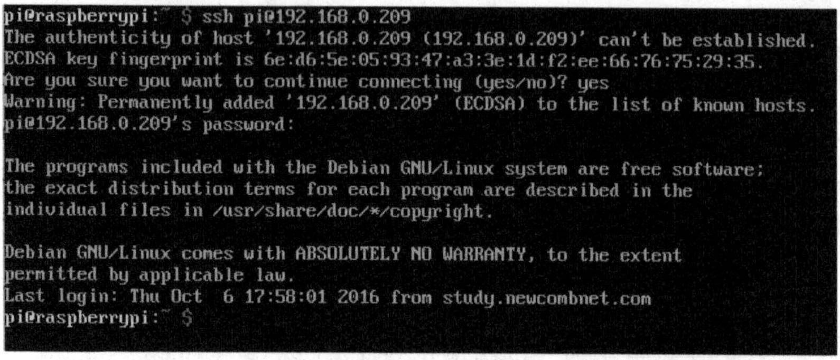

Figure 5-17. *Connecting to a Raspberry Pi using ssh on Linux*

As on the Macintosh, if this is the first time you've connected to this remote system, you'll be prompted to accept the SSH server's encryption key. Type **yes** and press the Enter key to continue. Once you're connected, you'll be at the prompt and ready to start entering commands.

Android/iPhone

There are many SSH client apps available for smartphones. For Android, I recommend ConnectBot, which is a free app that allows for multiple saved connections (see Figure 5-18). You can download ConnectBot from the Google Play Store.

Figure 5-18. *Using ConnectBot on Android to connect to a Raspberry Pi*

For iPhone, I recommend Cathode, which costs $4.99 and emulates the look and feel of classic hardware terminals (see Figure 5-19). You can download Cathode from the Apple App Store.

Figure 5-19. *Cathode running on an Apple iPhone*

Remote Desktops: vnc

So, you want to connect to your Raspberry Pi to use the desktop rather than using the command line? No problem. You can do this by using *Virtual Network Computing* (VNC) tools. Again, this requires two components: a VNC viewer on your local computer and a VNC server on the remote system. The latest version of Raspbian comes with the VNC server and viewer software installed. I won't cover installation, but will show you how to configure the server and client software necessary to view your Linux desktop remotely.

Setting Up the Raspberry Pi

In order to view the desktop remotely, you must first set up the VNC server on the Raspberry Pi itself. You can do this from the console or terminal emulator locally, or even via ssh remotely. All you need to do is enable it and then start it up.

Enable the VNC Server software by running the following command:

```
sudo systemctl enable vncserver-x11-serviced.service
```

Then start the VNC Server by running the following command (see Figure 5-20):

```
sudo systemctl start vncserver-x11-serviced.service
```

The VNC Server will now start automatically every time you boot up your Raspberry Pi.

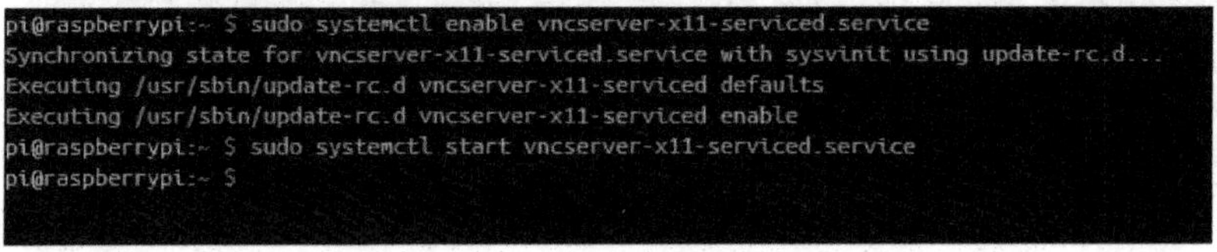

```
pi@raspberrypi:~ $ sudo systemctl enable vncserver-x11-serviced.service
Synchronizing state for vncserver-x11-serviced.service with sysvinit using update-rc.d...
Executing /usr/sbin/update-rc.d vncserver-x11-serviced defaults
Executing /usr/sbin/update-rc.d vncserver-x11-serviced enable
pi@raspberrypi:~ $ sudo systemctl start vncserver-x11-serviced.service
pi@raspberrypi:~ $
```

Figure 5-20. *Enabling and starting the VNC server on the Raspberry Pi*

The Raspberry Pi auto-senses the display it's connected to. When you don't have a display connected anymore, it will default to the lowest resolution possible, which is very small indeed. So in order to use VNC without a monitor attached, you will need to edit some configuration settings to tell your Raspberry Pi to default to a bigger screen size. Connect to your Raspberry Pi and edit the */boot/config.txt* file on the command line by typing:

```
sudo nano /boot/config.txt
```

Scroll down to the bottom of the file and add the following lines (see Figure 5-21):

```
hdmi_force_hotplug=1
hdmi_ignore_edid=0xa5000080
hdmi_group=2
hdmi_mode=16
```

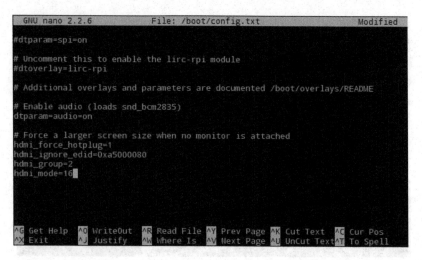

```
  GNU nano 2.2.6          File: /boot/config.txt              Modified

#dtparam=spi=on

# Uncomment this to enable the lirc-rpi module
#dtoverlay=lirc-rpi

# Additional overlays and parameters are documented /boot/overlays/README

# Enable audio (loads snd_bcm2835)
dtparam=audio=on

# Force a larger screen size when no monitor is attached
hdmi_force_hotplug=1
hdmi_ignore_edid=0xa5000080
hdmi_group=2
hdmi_mode=16█




^G Get Help  ^O WriteOut  ^R Read File ^Y Prev Page ^K Cut Text  ^C Cur Pos
^X Exit      ^J Justify   ^W Where Is  ^V Next Page ^U UnCut Text^T To Spell
```

Figure 5-21. *Editing the config.txt file on the Raspberry Pi*

The hdmi_force_hotplug setting tells your Pi that an HDMI display is attached, and the hdmi_mode setting forces a resolution of 1024×768 at 60Hz.

This should give you a workable desktop area even when no monitor is attached. Press Ctrl-X, then Y, then Enter to save your file and quit nano. You will need to reboot your Raspberry Pi for the changes to take effect.

Windows

Download and install the RealVNC software (*http://www.realvnc.com*). If you installed VNC Server, you will also need to register for a free personal use license as part of the install process. You can choose during install whether or not you want to install the server and viewer or just the viewer. We will only be using the viewer to connect to the Raspberry Pi. Once it's installed, launch the viewer from the Windows start menu (see Figure 5-22).

Figure 5-22. *Launching the VNC Viewer application on Windows*

When the VNC Viewer application starts, it will ask for the address of the remote system. Enter the IP address of your Raspberry Pi and click Connect (see Figure 5-23).

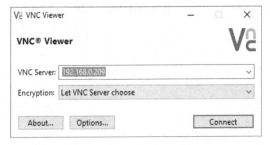

Figure 5-23. *Entering the IP address to connect to in VNC Viewer*

If this is the first time you've connected, you will be shown the server's unique signature and asked if you want to proceed. After you accept, you'll be asked for your Raspberry Pi username and password (see Figure 5-24).

Figure 5-24. *Entering your Raspberry Pi username and password in VNC Viewer*

After you click OK, you should be presented with a window showing your Raspberry Pi desktop.

MacOS

Download and install the RealVNC software (*https://www.realvnc.com*). If you installed VNC Server, you'll also need to register for a free personal use license as part of the install process. You can choose during install whether or not you want to install the server and viewer or just the viewer. We will only be using the viewer to connect to the Raspberry Pi. Once it's installed, launch the viewer by opening Finder and navigating to Applications→RealVNC→VNC Viewer (see Figure 5-25).

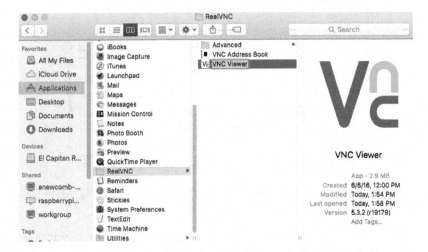

Figure 5-25. *Launching the VNC Viewer application on macOS*

When the VNC Viewer application starts, it will ask for the address of the remote system. Enter the IP address of your Raspberry Pi and click Connect (see Figure 5-26).

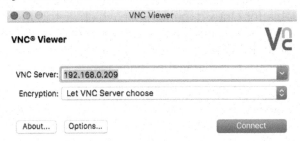

Figure 5-26. *Entering the IP address to connect to in VNC Viewer*

If this is the first time you've connected, you will be shown the server's unique signature and asked if you want to proceed. After you accept, you will be asked for your Raspberry Pi username and password (see Figure 5-27).

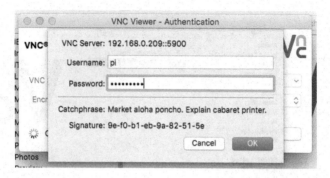

Figure 5-27. *Entering your Raspberry Pi username and pass-word in VNC Viewer*

After you click OK, you should be presented with a window showing your Raspberry Pi desktop.

Linux

Download and install the RealVNC software (*https:// www.realvnc.com/download/vnc/*). RealVNC has many pack-ages available for Linux depending on which distribution you are running. If you are running a Debian- or RedHat-based system, there are preconfigured packages available. Otherwise, you can install a general Linux installation package.

If you installed VNC Server, you'll also need to register for a free personal use license as part of the install process. You can choose during install whether or not you want to install the server and viewer or just the viewer. We will only be using the viewer to connect to the Raspberry Pi. Once it's installed, launch the viewer from the menu in the taskbar (see Figure 5-28).

Figure 5-28. *Launching the VNC Viewer application on Linux Mint*

You can also launch the VNC Viewer from the command line:

```
vncviewer
```

Once you launch VNC Viewer, you will need to accept the EULA before you can continue. The process from here is similar to Windows and macOS. Enter the IP address of your Raspberry Pi and click Connect (see Figure 5-29).

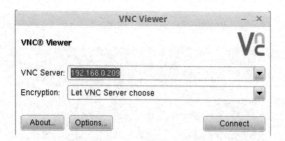

Figure 5-29. *Entering the IP address to connect to in VNC Viewer*

If this is the first time you've connected, you'll be shown the server's unique signature and asked if you want to proceed. After you accept, you will be asked for your Raspberry Pi username and password (see Figure 5-30).

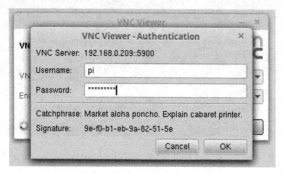

Figure 5-30. *Entering your Raspberry Pi username and password in VNC Viewer*

After you click OK, you should be presented with a window showing your Raspberry Pi desktop.

Android/iPhone

You can also access your Raspberry Pi desktop from your smartphone or tablet. There are versions of VNC Viewer for Android on the Google Play Store or for iPhone on the Apple App Store. The method for connecting is very similar to the desktop versions of VNC Viewer. Depending on the size of your display, however, you may find it frustrating to use on a small screen (see Figure 5-31).

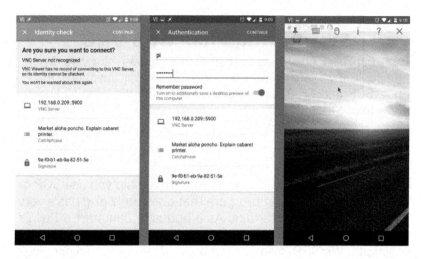

Figure 5-31. *Connecting to the Raspberry Pi with the mobile VNC Viewer app*

Transferring Files: scp, sftp

There are many ways to get files to and from an SBC running Linux. However, putting files onto the SD card directly is not easy from Windows and macOS. Since the SD card uses a Linux-based filesystem for primary storage, that part of the card won't be visible when you plug it into either of those systems. You could mount a network drive or use a USB pen drive to transfer the files, but these processes are cumbersome and time-consuming. This process is made more difficult when you're running headless since you can't see the desktop or have direct access to the console.

Luckily, there are easy tools built into Linux that help when you are transferring a few files over a network. Secure Copy (SCP) and Secure File Transfer Protocol (SFTP) use SSH to transfer files to a remote machine securely. scp is best used to transfer a single item like a file or an entire directory, while sftp can be used like regular FTP to create new directories and move a select group of files. This is one time that a graphical client is probably easier to use than the command line, but I will show you both ways where applicable.

Tranferring Files with VNC

You can also copy files with the VNC Viewer software from RealVNC if you are running the desktop on your Raspberry Pi. If not, SCP and SFTP should always work, so it's good to know how to use them.

Windows

In Windows, you will will need a program to help you use SCP or SFTP. WinSCP is a great program that does both, and has a very nice drag-and-drop interface. As a bonus, it can start a PuTTY session for you if you already have PuTTY installed. You can download WinSCP from its website (*https://winscp.net/eng/download.php*).

Once you have WinSCP installed, open it and choose an interface style. Personally, I like the Commander style, as it makes it easy to drag files back and forth between your computer and Raspberry Pi. Next, you will be presented with the login screen (see Figure 5-32).

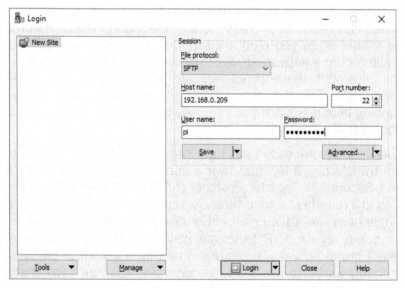

Figure 5-32. *Connecting to the Raspberry Pi with WinSCP*

Leave the "File protocol" set to SFTP (SFTP and SCP will work the same way with this program) and enter your Raspberry Pi's IP address in the "Host name" box. Next, fill in the "User name" and "Password" boxes with the username and password for your Raspberry Pi. Click the Login button to connect to your remote system. If this is the first time you've connected with WinSCP, you will be asked to verify the server's encryption key. WinSCP will then connect automatically using the credentials you supplied, and you will see the interface you picked (see Figure 5-33).

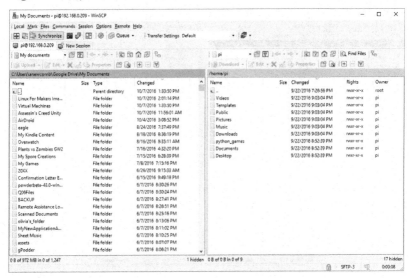

Figure 5-33. *The file browser window in WinSCP*

You can now transfer files back and forth between your computer and your Raspberry Pi.

MacOS

There are many clients available on macOS for SCP and SFTP file transfer. However, some of them include advanced functionality and can be quite expensive. You can find free tools in the Apple App Store if all you want to do is transfer files to your Raspberry Pi or other SBC. One such tool is Commander One. The free version of Commander One offers a Commander-style view of your files and will connect easily to other computers using many protocols. You can download Commander One from the developer's website (*http://mac.eltima.com/file-manager.html*) or in the Apple App Store.

Once you have Commander One installed, open it to find the default view of your local files. To open an SFTP session to your Raspberry Pi, click on the Connections Manager icon (see Figure 5-34).

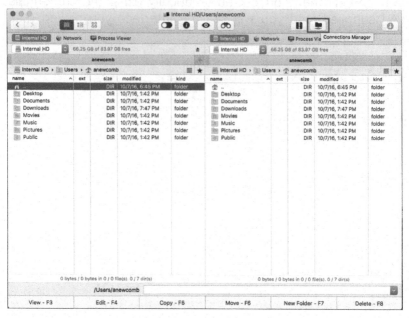

Figure 5-34. *Default file manager view in Commander One*

The Connections Manager window will allow you to use various methods to connect to remote systems. Click on the SFTP button to create a new SFTP connection (see Figure 5-35).

In the New Connection screen that appears, give this connection a name. Then, fill in the Raspberry Pi's IP address, your username, and password. Then click the Connect button (see Figure 5-36).

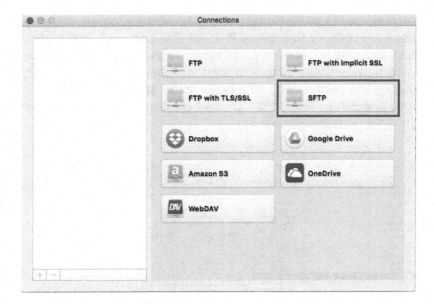

Figure 5-35. *Choosing a protocol in the Connections Manger screen of Commander One*

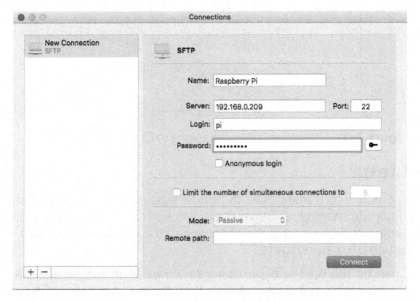

Figure 5-36. *Connection settings in Commander One*

Commander One will replace one pane of the program with the filesystem of your Raspberry Pi. It will also add a link to this connection to the top of each pane. Now you can drag files back and forth between your Mac and your Raspberry Pi. You can also change the permissions on a file or folder by right-clicking on it and choosing "Get info" or by selecting it and pressing Command-I (see Figure 5-37).

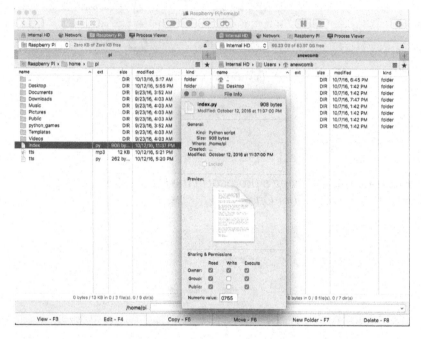

Figure 5-37. *Changing file permissions in Commander One*

Linux

The ability to transfer files to remote systems is built into most Linux file managers. Since there are so many distributions of Linux, I will be using Linux Mint, one of the most popular distributions, to demonstrate.

Open up the built-in file browser (in Linux Mint, the default file browser is Nemo), and click File, then "Connect to Server" (see Figure 5-38).

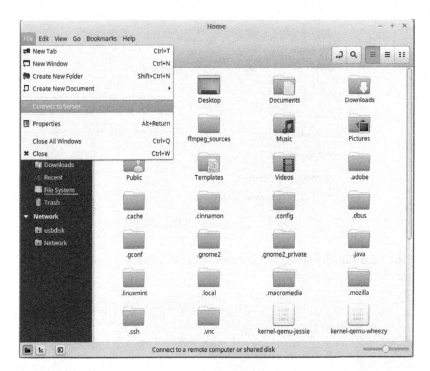

Figure 5-38. *Adding a new connection to a server in the Linux Mint file browser*

In the Type drop-down box, select SSH. Fill in the Raspberry Pi's IP address, username, and password and click the Connect button (see Figure 5-39).

If this is the first time you've connected to your Raspberry Pi, you'll be asked to verify that you want to make the connection. Click on Log In Anyway to continue. The file browser will open a new window displaying the filesystem of your Raspberry Pi. It will also add an icon under your network connections that represents this connection so you can easily get back to it at any time. If you want to end the connection, you can click the eject icon next to the name of the connection (see Figure 5-40).

Figure 5-39. *Entering the connection details for a new connection to the Raspberry Pi*

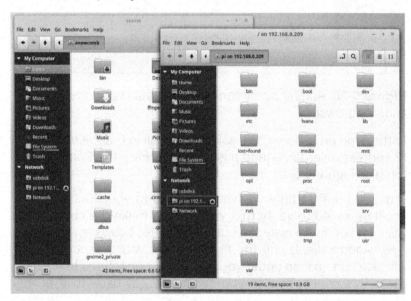

Figure 5-40. *A new window showing the files on the Raspberry Pi*

From the Command Line: MacOS and Linux

Both macOS and Linux have built-in command-line tools that let you use SCP and/or SFTP. Using SCP on the command line is an easy way to transfer a single file or directory to your Raspberry Pi. Using SFTP is a little more complex in that you will need to know how FTP commands work. Since command-line SFTP is probably not going to be used that often, I will only cover SCP in this section.

Open up the Terminal program in macOS or a terminal session on Linux. The scp command syntax is similar to ssh. To transfer a file to your home directory on your Raspberry Pi, just use the scp command by typing:

```
scp myfile username@IPaddress:/home/pi
```

where:

- *myfile* is the name of the file you want to transfer
- *username* is the username on your remote system
- *IPaddress* is your remote system's IP address

Be sure to include a space between your filename and the username of your remote system. After you enter your password, you will see a progress indication while the file is transferred (see Figure 5-41).

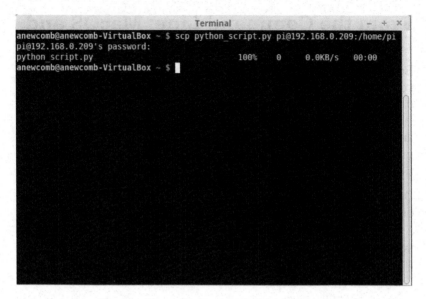

Figure 5-41. *Tranferring a file using scp on the Linux command line*

Why This Matters for Makers

More often than not, I find that Makers want to use the Raspberry Pi or some other SBC in a project where leaving a monitor and keyboard connected at all times is not a practical solution. Robots, security cameras, and LED light displays are all fun projects, but they are best when run headless. Knowing how to communicate with and control your Raspberry Pi remotely will open up a whole new world of possibilities for you to discover.

6/Tips and Tricks

Now that you can use the command line effectively from any-where, I'll explain some of the functionality an efficient Maker should know when putting their programs to work for them. These are the topics that come up again and again in forums and in conversations with new Linux users that a system admin-istrator would know to do like the back of their hand. This is by no means an exhaustive list, but by learning these tips and tricks, you can save a lot of time when building your projects and impress your friends by demonstrating your mad Linux command-line skillz.

Changing Your Hostname

By default, the hostname of a Raspberry Pi running Rasbian is raspberrypi. If you have more than one Raspberry Pi on your network, it can be confusing to know one from the other. So it is very helpful to have different hostnames for each Raspberry Pi on your network. You can change the hostname to be whatever you want in just a few simple steps.

 Don't Be Afraid of Change

Changing the hostname will not affect your Rasp-berry Pi's IP address. The only thing that changes is the name that shows up in network discovery tools like Fing or your router configuration website.

First, verify your existing hostname by running the `hostname` command:

```
hostname
```

This command without any options will only display the host-name and will not change anything (see Figure 6-1).

```
pi@raspberrypi:~ $ hostname
raspberrypi
pi@raspberrypi:~ $ _
```

Figure 6-1. *Running the hostname command to check the hostname*

Second, you will need to edit your *hosts* file. This special file is like a personal map that Linux uses to relate hostnames to IP addresses, and it supersedes information that might come from other devices on the network. Open the file for editing with **sudo** and replace all occurrences of **raspberrypi** with whatever you want your new hostname to be (see Figure 6-2):

```
sudo nano /etc/hosts
```

You can see in Figures 6-2 and 6-3 that I have replaced **raspber rypi** with **virtualpi**.

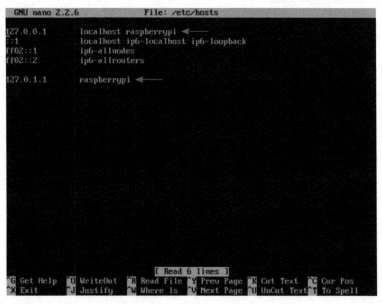

Figure 6-2. *Editing the hosts file*

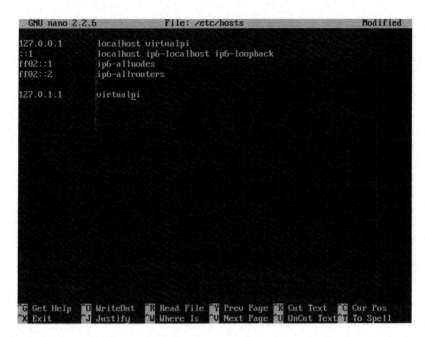

Figure 6-3. *Host file with new hostname*

The last step is to edit the *hostname* file. This file only contains the name of your system. Open the file for editing with **sudo** and replace **raspberrypi** with the same name you just used in the *hosts* file (see Figure 6-4):

```
sudo nano /etc/hostname
```

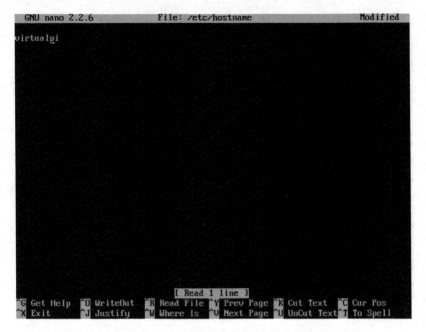

Figure 6-4. *Editing the hostname file*

Once these steps are complete, reboot your Raspberry Pi to make sure all the programs that use the hostname are using the new one. Now your Raspberry Pi should show up in network discovery tools with the new hostname. You will also see the new hostname at the prompt (see Figure 6-5). You can run the host name command again to verify your hostname at any time.

 Make Sure They Match

If the names in the *hosts* file and the *hostname* file do not match, you may end up having trouble connecting to your system over the network. If this happens, you will have to connect to the Raspberry Pi directly in order to fix it.

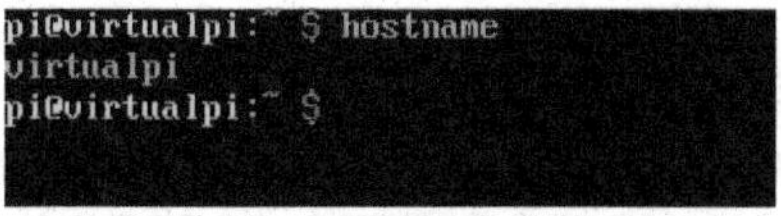

Figure 6-5. *The hostname command displaying the new hostname*

Then check your network discovery tools to see the new hostname on your network.

Starting a Script on Bootup: rc.local

Inevitably, you will want to run a script or program automatically when your Raspberry Pi boots up. This is especially important when you are running in headless mode and want your project to come alive all by itself when it's plugged in. The easiest way to do this is to add the script or program to a file called *rc.local*, which is located in the */etc* directory. All you need to do is edit that file using **sudo** and insert a line that runs your script or program. You should add your program just before the **exit 0** statement (see Figure 6-6):

```
sudo nano /etc/rc.local
```

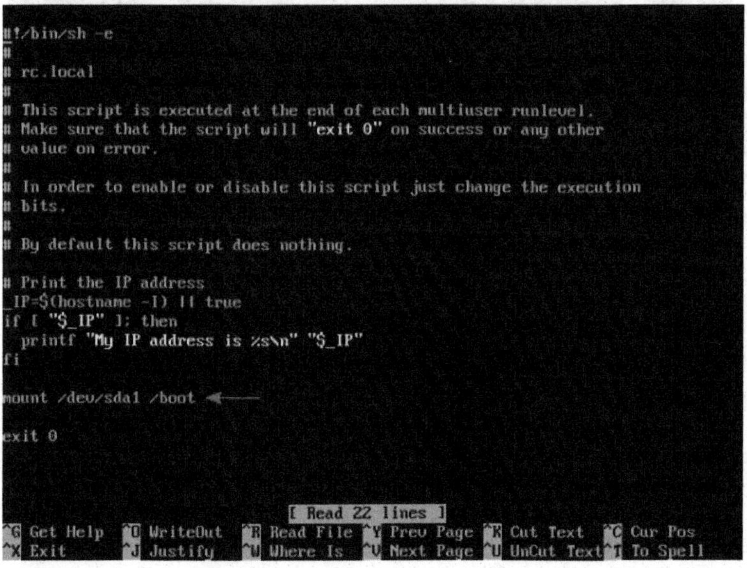

Figure 6-6. *Changing the rc.local file*

No Need for sudo

The *rc.local* script is run as "root," so you don't need to use sudo to run commands the way you normally might when you are logged in as the "pi" user. However, you will need to use the full path to your script or program.

Try It for Yourself

Add the *hello.sh* script to *rc.local* so it will run automatically when the Raspberry Pi boots up. Start by editing the *rc.local* file:

```
sudo nano /etc/rc.local
```

Now add the command we used in Chapter 2 to run the script, but use the full path to the file (see Figure 6-7).

```
  GNU nano 2.2.6           File: /etc/rc.local                    Modified

#!/bin/sh -e
#
# rc.local
#
# This script is executed at the end of each multiuser runlevel.
# Make sure that the script will "exit 0" on success or any other
# value on error.
#
# In order to enable or disable this script just change the execution
# bits.
#
# By default this script does nothing.

# Print the IP address
_IP=$(hostname -I) || true
if [ "$_IP" ]; then
  printf "My IP address is %s\n" "$_IP"
fi

sh /home/pi/hello.sh  ◄────

exit 0




^G Get Help  ^O WriteOut  ^R Read File  ^Y Prev Page  ^K Cut Text   ^C Cur Pos
^X Exit      ^J Justify   ^W Where Is   ^V Next Page  ^U UnCut Text ^T To Spell
```

Figure 6-7. *Using the rc.local file to run a script or program at boot time*

Save the file and exit by pressing Ctrl-X, then Y, then Enter. Now reboot your system with the command:

```
sudo shutdown -r now
```

When you system reboots, look for the "Hello World" print statements near the end of the boot process (see Figure 6-8).

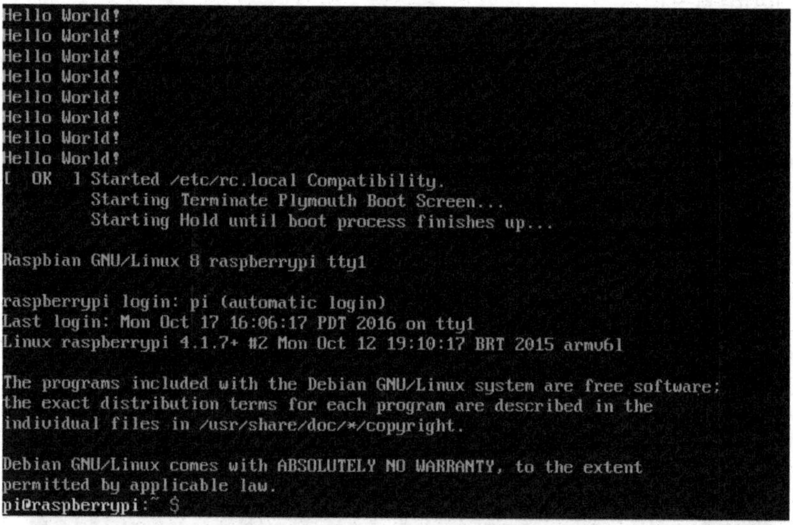

Figure 6-8. *The hello.sh script running at boot time*

Aliases

An alias in Linux is a way to tell the shell, "When I type this thing,
I want you to actually do this other thing." For example, when
you type ls on your Raspberry Pi, the shell is actually executing
ls --color=auto. This is because most terminals support color,
but by default the ls command does not turn the color option
on. Typing --color=auto every time would be a huge inconven-
ience, so there is an alias to handle that for you.

Aliases in Linux are handled mostly by running a script
called *.bashrc* when you log into the system. Each user has their
own *.bashrc* file located in their home directory. So to set up a
custom command or to change the default way an existing command works, you need to edit this file to set up your own aliases.
From the console or terminal emulator, edit the *.bashrc* file by
typing:

```
nano .bashrc
```

Be careful editing this file, as it is full of a lot of settings and configuration information for your shell. Scroll down until you see a section that defines the aliases for your session (see Figure 6-9).

```
  GNU nano 2.2.6              File: .bashrc

# If this is an xterm set the title to user@host:dir
case "$TERM" in
xterm*|rxvt*)
    PS1="\[\e]0;${debian_chroot:+($debian_chroot)}\u@\h: \w\a\]$PS1"
    ;;
*)
    ;;
esac

# enable color support of ls and also add handy aliases
if [ -x /usr/bin/dircolors ]; then
    test -r ~/.dircolors && eval "$(dircolors -b ~/.dircolors)" || eval "$(dirc$
    alias ls='ls --color=auto'
    #alias dir='dir --color=auto'
    #alias vdir='vdir --color=auto'

    alias grep='grep --color=auto'
    alias fgrep='fgrep --color=auto'
    alias egrep='egrep --color=auto'
fi

# colored GCC warnings and errors
#export GCC_COLORS='error=01;31:warning=01;35:note=01;36:caret=01;32:locus=01:q$

# some more ls aliases

^G Get Help    ^O WriteOut    ^R Read File   ^Y Prev Page   ^K Cut Text    ^C Cur Pos
^X Exit        ^J Justify     ^W Where Is    ^V Next Page   ^U UnCut Text  ^T To Spell
```

Figure 6-9. *Editing the .bashrc file*

A good place to add your aliases is after the fi statement in this section, as it keeps them all together, but technically it doesn't matter. Another good place to add statements to a file like this is at the end of the file so that they are easy to find. Syntax is important here, so be sure to make the name of your new alias all one word with no spaces (i.e., "runthis" and not "run this"). Also, make sure there are no spaces before or after the = sign.

 Log Out to Apply Changes

The *.bashrc* file is only read by the system when you log in, so in order to apply any changes you make to that file you need to log out and log in again by using the exit command.

Try It for Yourself

Open the *.bashrc* file for editing:

```
nano .bashrc
```

Add an alias called `lsbydate` that will sort the output of the `ls` command by last modified date in ascending order like so (see Figure 6-10):

```
alias lsbydate='ls -ltr'
```

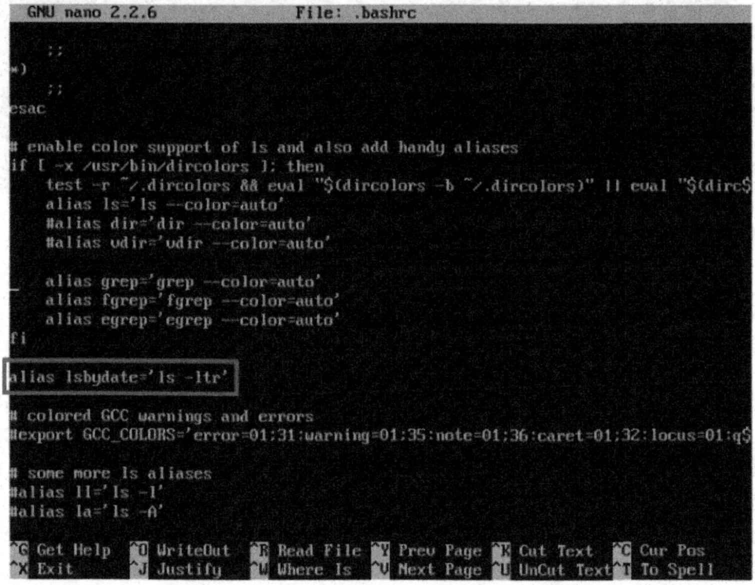

Figure 6-10. *Adding the lsbydate alias to the .bashrc file*

Save the file and exit by pressing Ctrl-X, then Y, then Enter. Now exit out of your session by typing:

```
exit
```

Log back in as the same user and try running your command. You should see the files and directories in your current location sorted by date (see Figure 6-11).

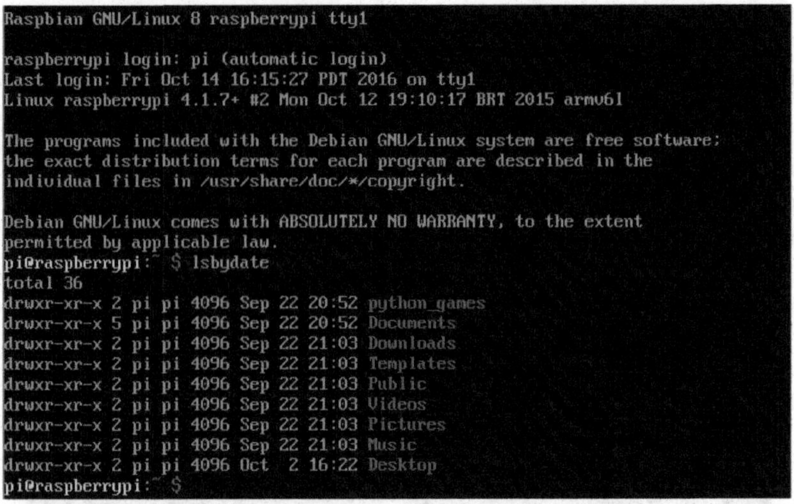

Figure 6-11. *Using the new lsbydate alias*

Checking Disk and File Space Usage: df, du

Although you can add extra storage via the USB connections on your Raspberry Pi, your primary storage is your SD card. Since SD card storage is rather limited in size, you will most likely want to know how much of your storage has been used and how much is still available. You can easily see how much space you have on your SD card (or any other mounted filesystem) by using the df utility, which stands for *disk filesystem*.

The df utility shows you a list of all the mounted filesystem devices, their total size, how much space is used, how much is available, the percentage of space used, and where the filesystem is mounted. The important filesystems to keep track of are the ones mounted on / and /boot because they represent your primary storage and your boot partition, respectively.

By default, the output of df is formatted to display kibibytes (1,024 bytes), but you can change this to display more human-readable output by using the -h option (see Figure 6-12).

```
pi@raspberrypi:~ $ df
Filesystem     1K-blocks    Used Available Use% Mounted on
/dev/root      15180208 4227616  10275516  30% /
devtmpfs         469536       0    469536   0% /dev
tmpfs            473868       0    473868   0% /dev/shm
tmpfs            473868    6428    467440   2% /run
tmpfs              5120       4      5116   1% /run/lock
tmpfs            473868       0    473868   0% /sys/fs/cgroup
/dev/mmcblk0p1    64456   21192     43264  33% /boot
tmpfs             94776       4     94772   1% /run/user/1000
pi@raspberrypi:~ $ df -h
Filesystem     Size  Used Avail Use% Mounted on
/dev/root      15G  4.1G  9.8G  30% /
devtmpfs       459M    0  459M   0% /dev
tmpfs          463M    0  463M   0% /dev/shm
tmpfs          463M  6.3M  457M   2% /run
tmpfs          5.0M  4.0K  5.0M   1% /run/lock
tmpfs          463M    0  463M   0% /sys/fs/cgroup
/dev/mmcblk0p1  63M   21M   43M  33% /boot
tmpfs           93M  4.0K   93M   1% /run/user/1000
pi@raspberrypi:~ $ ▊
```

Figure 6-12. *The output of the df command*

You might also want to know how much space a particular file or directory is occupying on your filesystem. You could use the ls command to get a list of all the files in your current directory and add them all up, but it is easier to use the du tool, which stands for *disk usage*. By default, the du tool shows the size of every file and directory starting from your current location and proceeding recursively through the filesystem until there is nothing more to show.

Like df, du will display the sizes in kibibytes unless you use the -h option to show human-readable sizes. You can also specify the number of subdirectories from your current location about which you want to display detailed information with the -d option. For example, to show only the summary of your current directory, you would use the option -d 0. If you want to also see the summary for just your directory and all immediate subdirectories, you would use the option -d 1 (see Figure 6-13).

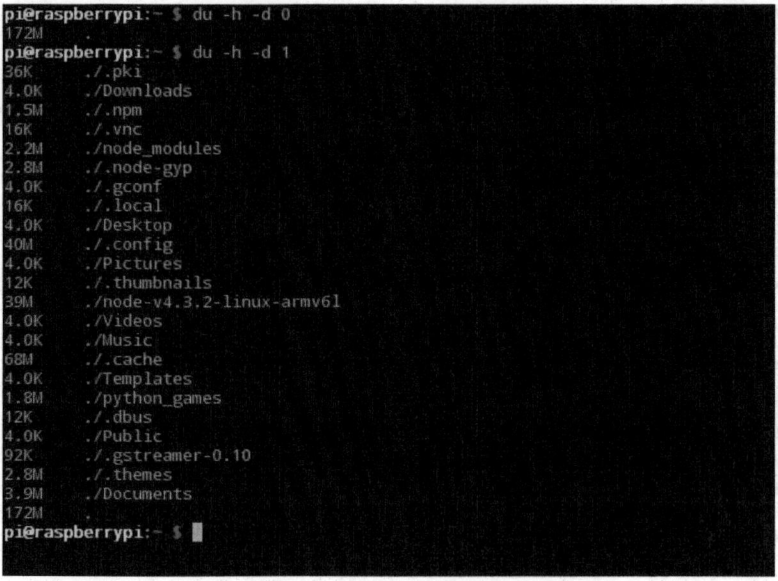

```
pi@raspberrypi:~ $ du -h -d 0
172M    .
pi@raspberrypi:~ $ du -h -d 1
36K     ./.pki
4.0K    ./Downloads
1.5M    ./.npm
16K     ./.vnc
2.2M    ./node_modules
2.8M    ./.node-gyp
4.0K    ./.gconf
16K     ./.local
4.0K    ./Desktop
40M     ./.config
4.0K    ./Pictures
12K     ./.thumbnails
39M     ./node-v4.3.2-linux-armv6l
4.0K    ./Videos
4.0K    ./Music
68M     ./.cache
4.0K    ./Templates
1.8M    ./python_games
12K     ./.dbus
4.0K    ./Public
92K     ./.gstreamer-0.10
2.8M    ./.themes
3.9M    ./Documents
172M    .
pi@raspberrypi:~ $ █
```

Figure 6-13. *The output of the du command*

In Figure 6-12, you can see that the total amount of disk space my current location (*/home/pi*) is taking up is 172 MB, whereas my *Documents* subdirectory is taking up 3.9 MB.

Performance Monitoring: top

There are whole books dedicated to the topic of Linux performance. Instead of covering everything possible, I will just touch on the basics here that a Maker should know. Once you've discovered all the project possibilities that can happen with an SBC like the Raspberry Pi, you might be tempted to do a lot of different things with it at the same time. Indeed, that is one of the benefits of an SBC over a microcontroller platform like Arduino —the Pi can read sensors and drive motors and send tweets, almost simultaneously.

However, you can push things too far and start running out of resources. Your project may slow to a crawl, or even crash completely. It is important to be able to monitor the performance of your system so you can shut down unimportant functions if things start slowing down. You can monitor CPU utilization right from the desktop, as there is a CPU percentage indicator applet right in the taskbar that shows total CPU usage at that particular point in time (see Figure 6-14).

Figure 6-14. *The CPU performance applet on the Raspberry Pi desktop*

Unfortunately, this applet doesn't show memory or storage I/O utilization and, of course, it is only visible when you are running the desktop. So when you want more detailed information, it is good to use the **top** tool. **top** stands for *table of processes*, and as the name suggests, it lists running processes in table form sorted by their resource utilization. By default, the displayed information is refreshed every three seconds and is sorted by CPU utilization. **top** displays a large amount of information at one time, so let's take a look at a breakdown of what it all means so you can use it to monitor or debug your project (see Figure 6-15).

```
top - 19:38:30 up 3 days, 20:29,  2 users,  load average: 0.62, 0.20, 0.06
Tasks: 157 total,   1 running, 156 sleeping,   0 stopped,   0 zombie
%Cpu(s):  1.2 us,  0.6 sy,  0.0 ni, 98.2 id,  0.0 wa,  0.0 hi,  0.0 si,  0.0 st
KiB Mem:  947740 total,  606868 used,  340872 free,   78044 buffers
KiB Swap: 102396 total,       0 used,  102396 free,  320672 cached Mem

  PID USER      PR  NI    VIRT    RES    SHR S  %CPU %MEM     TIME+ COMMAND
  522 root      20   0   33412  17280  10408 S   2.6  1.8  51:47.17 vncserver-x11-c
19491 pi        20   0  416920 102676  51436 S   2.0 10.8   0:09.07 chromium-browse
  651 root      20   0  166780  52980  35516 S   1.0  5.6   3:20.74 Xorg
19523 pi        20   0    5112   2476   2092 R   1.0  0.3   0:00.22 top
   79 root     -51   0       0      0      0 S   0.3  0.0   1:42.23 irq/92-mmc1
   85 root      20   0       0      0      0 S   0.3  0.0   0:05.26 mmcqd/0
  673 root      20   0   10940   7204   6736 S   0.3  0.8   0:10.28 vncagent
 1104 pi        20   0   47220  19012  16060 S   0.3  2.0   0:03.57 lxterminal
19259 root      20   0       0      0      0 S   0.3  0.0   0:00.46 kworker/u8:1
    1 root      20   0   23892   3968   2740 S   0.0  0.4   0:11.32 systemd
    2 root      20   0       0      0      0 S   0.0  0.0   0:00.12 kthreadd
    3 root      20   0       0      0      0 S   0.0  0.0   0:00.92 ksoftirqd/0
    5 root       0 -20       0      0      0 S   0.0  0.0   0:00.00 kworker/0:0H
    7 root      20   0       0      0      0 S   0.0  0.0   0:54.26 rcu_sched
    8 root      20   0       0      0      0 S   0.0  0.0   0:00.00 rcu_bh
    9 root      rt   0       0      0      0 S   0.0  0.0   0:00.10 migration/0
   10 root      rt   0       0      0      0 S   0.0  0.0   0:00.06 migration/1
```

Figure 6-15. *Running the top command*

Let's start with the topmost line (see Figure 6-16).

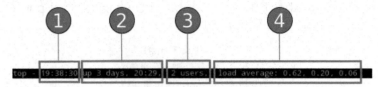

Figure 6-16. *Breakdown of the top command*

1. Current time
2. Uptime in days, hours, and minutes
3. Number of users logged in (if you are running the desktop, this will normally be 2)
4. Average CPU load over the last 1 minute, 5 minutes, and 15 minutes

---

 Uptime

You can also get this single line of information by running the command `uptime`.

Now let's look at the Tasks line (see Figure 6-17).

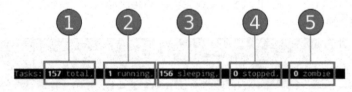

Figure 6-17. *Breakdown of the top command (continued)*

1. Total number of processes
2. Number of processes currently running
3. Number of processes currently idle
4. Number of processes that have received a stop signal (more on this later)
5. Number of processes that have exited but are waiting for another process to finish

Remember that in Chapter 2 you learned about parent and child processes. *Zombie* processes are usually child processes that have finished what they are doing but are required to wait for their parent process to exit before they can be cleared from the process list. Now let's examine the %CPU(s) line (see Figure 6-18).

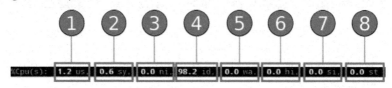

Figure 6-18. *Breakdown of the top command (continued)*

1. Percentage of time the total number of CPUs have spent running normal user processes. This is the key indicator of how busy the system is.
2. Percentage of time spent on running system kernel processes.
3. Percentage of time spent on running prioritized or de-prioritized processes. In Linux, this is referred to as *niceness*, but it isn't often used in small systems like the Raspberry Pi.
4. Percentage of time not doing anything, aka idle.
5. Percentage of time waiting for I/O to complete. This can be an indicator of using up all the memory or slow storage.
6. Percentage of time servicing hardware interrupts. This can happen when external devices need to send information to the CPUs right away.
7. Percentage of time servicing software interrupts. This is a less important kind of CPU signal than a hardware interrupt.
8. This only applies to virtualized systems and indicates the percentage of time stolen from the system because the host system was busy doing something else.

If you are having poor performance on your Raspberry Pi, it will normally show up as high percentage numbers in the user, system, or wait columns. Remember that the numbers in the %Cpu(s) row should add up to 100%. Now let's look at the KiB Mem and Kib Swap lines (see Figure 6-19).

Figure 6-19. *Breakdown of the top command (continued)*

1. Total amount of available memory and swap space in kibibytes
2. Amount of memory or swap space used
3. Amount of memory or swap space available
4. Amount of buffers and cached memory used

These two lines refer to different types of memory. *Mem* refers to physical memory and should be close to the amount of memory installed on the system. In this case, my Raspberry Pi 3 has 1 GB of RAM so the number shown in the total column should be close to that. *Swap* refers to space on a disk (in this case, the SD card) that is used as virtual memory just in case the system runs out of physical memory. Since the SD card is so much slower than physical memory, using swap will affect the performance of your Raspberry Pi.

Buffers refers to the memory used when the system mounts or accesses the filesystems connected to the system itself. The system keeps some of the information about the files and devices in memory to speed up repeated access. Cached memory refers to the actual data from the files and programs. As the data is read, it is loaded into memory and kept there for a period of time to make it faster to access. The system will move data in and out of memory automatically.

The key thing to be on the lookout for here is the amount of memory used. If you're constantly running out of memory, you should try and find what is causing it and close memory-hungry applications to free up some space. Now let's look at the last section of the screen that can help us do just that (see Figure 6-20).

```
  PID USER      PR  NI    VIRT    RES    SHR S  %CPU %MEM     TIME+ COMMAND
  522 root      20   0   33412  17280  10408 S   2.6  1.8  51:47.17 vncserver-x11-c
19491 pi        20   0  416920 102676  51436 S   2.0 10.8   0:09.07 chromium-browse
  651 root      20   0  166780  52980  35516 S   1.0  5.6   3:20.74 Xorg
19523 pi        20   0    5112   2476   2092 R   1.0  0.3   0:00.22 top
   79 root     -51   0       0      0      0 S   0.3  0.0   1:42.23 irq/92-mmc1
   85 root      20   0       0      0      0 S   0.3  0.0   0:05.26 mmcqd/0
  673 root      20   0   10940   7204   6736 S   0.3  0.8   0:10.28 vncagent
 1104 pi        20   0   47220  19012  16060 S   0.3  2.0   0:03.57 lxterminal
19259 root      20   0       0      0      0 S   0.3  0.0   0:00.46 kworker/u8:1
    1 root      20   0   23892   3968   2740 S   0.0  0.4   0:11.32 systemd
    2 root      20   0       0      0      0 S   0.0  0.0   0:00.12 kthreadd
    3 root      20   0       0      0      0 S   0.0  0.0   0:00.92 ksoftirqd/0
    5 root       0 -20       0      0      0 S   0.0  0.0   0:00.00 kworker/0:0H
    7 root      20   0       0      0      0 S   0.0  0.0   0:54.26 rcu_sched
    8 root      20   0       0      0      0 S   0.0  0.0   0:00.00 rcu_bh
    9 root      rt   0       0      0      0 S   0.0  0.0   0:00.10 migration/0
   10 root      rt   0       0      0      0 S   0.0  0.0   0:00.06 migration/1
```

Figure 6-20. *Breakdown of the top command (continued)*

PID
    The process ID.

USER
    The owner of the process. Usually the user that launched it.

PR
    The current priority of the process.

NI
    The *nice* value or user-defined priority of the process.

VIRT
    The total amount of memory needed by the process.

RES
    The amount of memory actually used by the process.

SHR
    The amount of shared memory available to a process.

S
    The current status of the process, which may be one of the
    following letters:

    • D = uninterruptible sleep
    • R = running
    • S = sleeping
    • T = traced or stopped
    • Z = zombie

%CPU
> The percentage of a single CPU that is being used by a process.

%MEM
> The percentage of total memory that this process is using.

TIME
> The amount of time that the CPUs have spent running this process, in hundredths of a second.

COMMAND
> The command used to launch the process. A + sign at the end of a command means that the command was too long to fit in this column.

The important things to keep track of in this part of the display are the processes that use a lot of CPU and memory. You can do this by watching the %CPU and %MEM columns. Usually any troublesome processes will quickly appear at the top of this list. Running processes will show up with all the columns highlighted for that line. You can also manipulate this list by using case-sensitive command keys.

x
> Highlight the current sort field.

P
> Sort by %CPU (default).

M
> Sort by %MEM.

N
> Sort by PID.

T
> Sort by TIME.

and
> Toggle sort by one column at a time left or right.

Arrow keys, PgUp, PgDown
> Scroll left, right, up, down.

k

    Kill a process.

h

    Help.

q

    Quit.

There are many other options with **top**. You can find out more by referencing the help screen or by reading the manpage.

Try It for Yourself

Find out how much resources a given app uses when it's started and after it's up and running. For now, this is most easily done from the desktop. Later, I will show you how to run a process in the background so you can do this from the command line. Open a terminal emulator window and launch **top**:

    top

Then start another application and watch the output of the **top** command to see how much resources are used in the first 30 seconds or so after the application is started (the web browser might be a good choice for this). Continue to watch to see what happens as the application finishes loading and is running without any activity. Start using the application (i.e., load a web page, open a file, etc.) to see what happens to CPU and memory while your application performs those functions.

Killing a Process: Ctrl-C, ps, kill

All operating systems have programs that get out of control in one way or another. Sometimes programs become unresponsive or there just isn't access to them to quit them through normal means. This could be because of a flaw in the program or operating system. In any case, you should know how to stop the program from running in order to prevent it from consuming too many resources and causing usability issues. In Linux, this is known as *killing* a program or process and is used to forcibly exit and terminate the program. For many programs, killing them doesn't cause any harm. However, in more complex programs, this could lead to program errors, as the program may not have a chance to clean up open files or network connections before it exits. If your program has an exit function built in (like a close button or an exit key), you should always try using that first before you resort to killing it.

By pressing Ctrl-C, you will send an interrupt signal to the program that's currently running in the terminal. In most cases, this will exit the program abruptly and return you to the prompt. In the case of a simple script, this can be the quickest way to exit and get on with the next task at hand.

In the case of jobs or programs that were started automatically or in another user session, you will first need to find out the PID of the process you want to kill. To do this, you can use the **ps** command, which stands for *process status*. When run without any options, the **ps** command will only show processes that your current user is running. To get a list of all processes, you can use the options -ef to get a complete list with more details (see Figure 6-21).

```
pi@raspberrypi:~ $ ps
  PID TTY          TIME CMD
 9593 pts/0    00:00:00 ps
29756 pts/0    00:00:00 bash
pi@raspberrypi:~ $ ps -ef
UID         PID  PPID  C STIME TTY          TIME CMD
root          1     0  0 20:42 ?        00:00:03 /sbin/init splash
root          2     0  0 20:42 ?        00:00:00 [kthreadd]
root          3     2  0 20:42 ?        00:00:00 [ksoftirqd/0]
root          5     2  0 20:42 ?        00:00:00 [kworker/0:0H]
root          7     2  0 20:42 ?        00:00:04 [rcu_sched]
root          8     2  0 20:42 ?        00:00:00 [rcu_bh]
root          9     2  0 20:42 ?        00:00:00 [migration/0]
root         10     2  0 20:42 ?        00:00:00 [migration/1]
root         11     2  0 20:42 ?        00:00:00 [ksoftirqd/1]
root         13     2  0 20:42 ?        00:00:00 [kworker/1:0H]
root         14     2  0 20:42 ?        00:00:00 [migration/2]
root         15     2  0 20:42 ?        00:00:00 [ksoftirqd/2]
root         17     2  0 20:42 ?        00:00:00 [kworker/2:0H]
root         18     2  0 20:42 ?        00:00:00 [migration/3]
root         19     2  0 20:42 ?        00:00:00 [ksoftirqd/3]
root         21     2  0 20:42 ?        00:00:00 [kworker/3:0H]
root         22     2  0 20:42 ?        00:00:00 [kdevtmpfs]
root         23     2  0 20:42 ?        00:00:00 [netns]
```

Figure 6-21. *The output of the ps command*

Try this for yourself and you'll soon realize that hundreds of processes can be listed in the output of this command. To find the process you're looking for, you can use **grep** to limit the results (see Figure 6-22). The **grep** command, when added to another command, prints out only the lines that match a given search string (more on grep later in this chapter):

```
ps -ef | grep search term
```

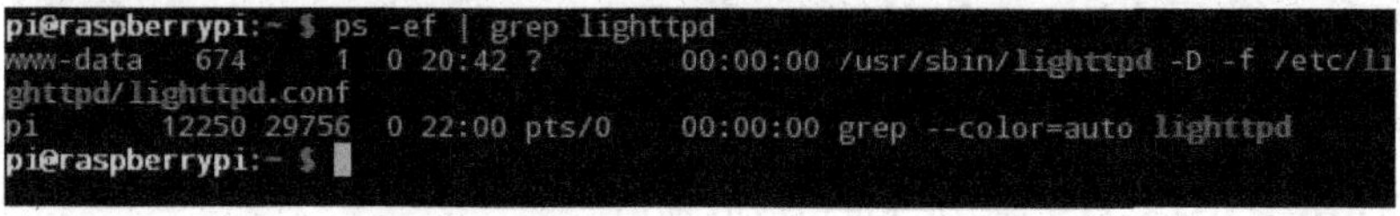

```
pi@raspberrypi:~ $ ps -ef | grep lighttpd
www-data   674     1  0 20:42 ?        00:00:00 /usr/sbin/lighttpd -D -f /etc/li
ghttpd/lighttpd.conf
pi       12250 29756  0 22:00 pts/0    00:00:00 grep --color=auto lighttpd
pi@raspberrypi:~ $
```

Figure 6-22. *Using grep with the ps command*

In this case, you can see that I searched for lighttpd, which represents the web server process I'm running. However, there is an extra result, which represents the grep search for lighttpd itself. You can ignore this result. The one we want is the top result, which in this instance has PID 674.

Let's suppose for a moment that my Lighttpd web server was locked up for some reason. Since it's a service, I've already tried to stop it with the proper command (sudo service lighttpd stop), but it didn't respond. In order to kill a process, you can use the kill command. The kill command sends a shutdown signal to the process and thus can stop a process abruptly. There are several options you can use with kill depending on how you want to end the process:

kill PID
 Send a normal terminate signal to the process.

kill -1 PID
 Send a restart signal to the process.

kill -2 PID
 Send an interrupt signal to the process. This is the same as pressing Ctrl-C.

kill -9 PID
 Send the kill signal and shut down the process immediately.

In this case, I want to end Lighttpd normally with a regular terminate signal. Because my current user didn't start that process, I will need to use sudo (see Figure 6-23).

```
pi@raspberrypi:~ $ ps -ef | grep lighttpd
www-data   674      1  0 20:42 ?        00:00:00 /usr/sbin/lighttpd -D -f /etc/li
ghttpd/lighttpd.conf
pi       24652 29756  0 22:21 pts/0    00:00:00 grep --color=auto lighttpd
pi@raspberrypi:~ $ sudo kill 674
pi@raspberrypi:~ $ ps -ef | grep lighttpd
pi        1032 29756  0 22:22 pts/0    00:00:00 grep --color=auto lighttpd
pi@raspberrypi:~ $ █
```

Figure 6-23. *Using the kill command*

After you send a kill command to a process, it's a good idea to run ps again to make sure it isn't running anymore. If a process does not respond to the kill command or a kill command with the -2 option, you can use the -9 option as a last resort. As you can see in Figure 6-23, after I tried to kill the lighttpd process and checked again, the only result I got back was the grep search itself, so the process was killed.

Stop, Background, and Foreground Jobs: Ctrl-Z, &, fg

Sometimes it can be helpful to pause a process, go do something, and then come back and continue where you left off. Other times, it might be nice to run a program in the background from the very beginning if you don't need to watch it the whole time. Linux has commands that can help you stop a job temporarily and run jobs in the background so they aren't in your way while you're working on the command line.

If you want to pause your program and come back to it later, you can use the Ctrl-Z keyboard shortcut. In Linux, this is known as *stopping* the program and will send your program to the background and return you to the prompt. The program will not process any more instructions until you bring it back to the foreground (see Figure 6-24).

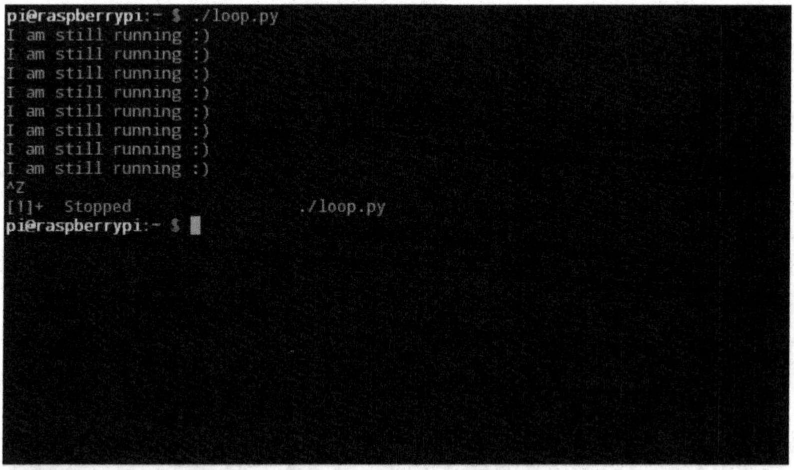

Figure 6-24. *Using Ctrl-Z to stop a process*

When you are ready to *unpause* the program, you can use the `fg` command. This will bring the program back to the screen and continue it from where it was stopped (see Figure 6-25).

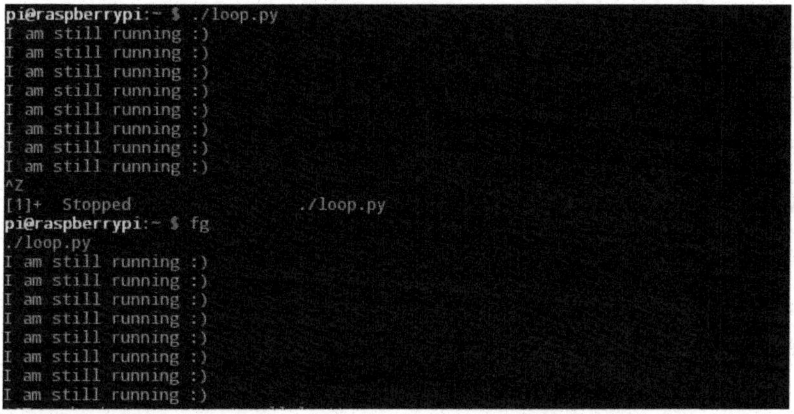

Figure 6-25. *Using fg to bring the process back to the foreground*

If you want to start a program running in the background when you launch it, you can add the & character to the end of the command. This will start the program and tell you its PID in case you need to reference it later, then return you to the prompt to do other tasks. However, if you have output coming from the program, it will still be printed on the screen, as you can see in Figure 6-26. Later, I will show you how to send this output somewhere else. To pull the program to the foreground, you can use the same fg command.

```
pi@raspberrypi:~ $ ./loop.py &
[1] 18089
pi@raspberrypi:~ $ I am still running :)
I am still running :)
I am still running :)
ls
dash2.py    loop.py                       omxplayer.log    python_script.py
dash.py     Music                         output_tts.mp3   Templates
Desktop     node_modules                  Pictures         tts.mp3
Documents   node-v4.3.2-linux-armv6l      parrot.py        tts.py
Downloads   node-v4.3.2-linux-armv6l.tar.gz  Public        Videos
index.py    nohup.out                     python_games
pi@raspberrypi:~ $ I am still running :)
I am still running :)
I am still running :)
fI am still running :)
g
./loop.py
I am still running :)
I am still running :)
I am still running :)
I am still running :)
I am still running :)
```

Figure 6-26. *Using & to run a program in the background*

Try It for Yourself

Create a looping script and practice running it in the background and bringing to the foreground. Find the PID with the ps command. Once you've found the PID, you can kill it with the kill command.

Open a new file called *loop.py*:

    nano loop.py

Now, copy the following text into the file:

    #!/usr/bin/python

    import time

    while True:

        print "I am still running :)"

        time.sleep(10)

Save the file by pressing Ctrl-X, then Y, then Enter. Run the program with the following command:

```
python loop.py
```

This script will print "I am still running :)" every 10 seconds. Stop it by pressing Ctrl-Z. You should be returned to the prompt, where you can enter some other commands. Bring the script back to the foreground by typing:

```
fg
```

Kill the script by using Ctrl-C. Run the script again, but start it in the background:

```
python loop.py &
```

This time, look up the PID of the script by searching for it with **ps** and **grep**.

```
ps -ef | grep loop
```

Kill the script process by using the associated PID.

```
kill PID
```

This should kill the process and stop it from printing those annoying messages. :)

Finding USB Devices: lsusb

Since most SBCs like the Raspberry Pi have *Universal Serial Bus* (USB) ports built in, using USB devices is an easy way to add functionality that your project requires. Keyboards, mice, audio devices, Bluetooth adapters, and WiFi adapters can all be connected via USB. Most Linux distributions support many current and legacy USB devices without requiring you to install any drivers. This is because the Linux kernel has the drivers already built in, thanks to the hard work of the many programmers who have contributed their code over the years.

However, typically you won't get a nicely formatted pop-up message telling you that your device has been recognized by the system, as you do with Windows. To get a list of USB devices currently recognized by your system, you can use the command lsusb. Similar to ls, this command lists your USB devices along with their hexadecimal device ID. It's a good idea to run this command once before you plug in your device and then again after you plug it in, as it's not always easy to tell one device from another (see Figure 6-27).

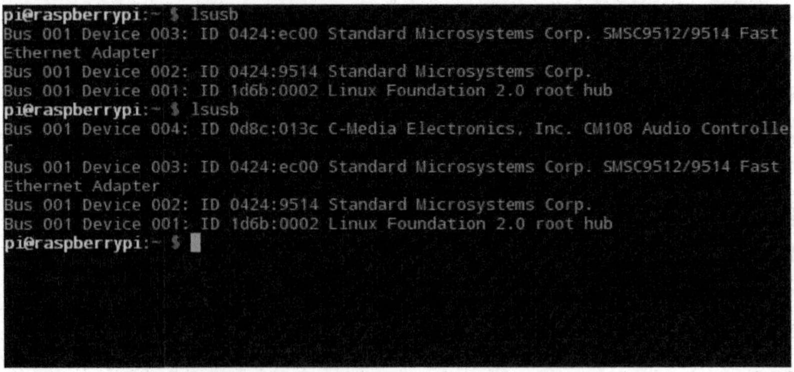

```
pi@raspberrypi:~ $ lsusb
Bus 001 Device 003: ID 0424:ec00 Standard Microsystems Corp. SMSC9512/9514 Fast
Ethernet Adapter
Bus 001 Device 002: ID 0424:9514 Standard Microsystems Corp.
Bus 001 Device 001: ID 1d6b:0002 Linux Foundation 2.0 root hub
pi@raspberrypi:~ $ lsusb
Bus 001 Device 004: ID 0d8c:013c C-Media Electronics, Inc. CM108 Audio Controlle
r
Bus 001 Device 003: ID 0424:ec00 Standard Microsystems Corp. SMSC9512/9514 Fast
Ethernet Adapter
Bus 001 Device 002: ID 0424:9514 Standard Microsystems Corp.
Bus 001 Device 001: ID 1d6b:0002 Linux Foundation 2.0 root hub
pi@raspberrypi:~ $
```

Figure 6-27. *The output of the lsusb command*

As you can see, before I plugged in any physical devices to my Raspberry Pi, I still had some USB devices listed when I ran lsusb. This is because some of the built-in devices like the Ethernet adapter are connected to the USB bus internally. Once I plugged in a USB microphone and ran lsusb again, I could see that my device showed up as "C-Media Electronics, Inc. CM108 Audio Controller." This information might be helpful to you if you need to look up information about your device on the internet. Also, some programs that you create may require you to reference the hexadecimal ID of the device in order to work properly.

Remember

Every device you plug into the USB ports on your Raspberry Pi will draw additional power. Drawing more power than is available will cause your system to crash, especially when you first boot it up. Be sure your power supply has enough current to supply both your Raspberry Pi and all the devices you plug into it.

Logging the Output of a Script: >, >>

There are several situations in which you might want to capture the output of a script. If you have a project that's collecting data over a long period of time, you will want to be able to capture that data so you can analyze it later. In some cases, you may be getting intermittent errors printing to the screen but don't have time to see them before they scroll off. If you're running a script at startup with *rc.local*, it will run as "root" and you won't see the output at all when you log in.

In all of these cases, you can capture the output of a script in a logfile that you can reference later to see what's going on. The way to do that in Linux is to modify the command that launches the script using the greater-than symbols. This will redirect the output.

>

Send the output to a new file.

>>

Append the output to an existing file.

&>, &>>

Create or append file including errors.

For example, if I wanted to send the output of the *loop.py* script we created earlier to a new file called *loop.log*, I could do it with the following command:

```
python loop.py > loop.log
```

While this script is running, all normal output (like print statements) will be written to the *loop.log* file. Once the script has exited, you can examine the logfile to see the output with a text editor or simply by using the `more` command (see Figure 6-28).

```
pi@raspberrypi:~ $ python loop.py > loop.log
^CTraceback (most recent call last):
  File "loop.py", line 7, in <module>
    time.sleep(1)
KeyboardInterrupt
pi@raspberrypi:~ $ more loop.log
I am still running :)
I am still running :)
I am still running :)
I am still running :)
I am still running :)
I am still running :)
I am still running :)
I am still running :)
I am still running :)
I am still running :)
I am still running :)
I am still running :)
pi@raspberrypi:~ $
```

Figure 6-28. *Using > to send output to a new file*

If you want to append to this same logfile without overwriting it, you need to use two greater-than symbols like this (see Figure 6-29):

```
python loop2.py >> loop.log
```

```
pi@raspberrypi:~ $ python loop2.py >> loop.log
^CTraceback (most recent call last):
  File "loop2.py", line 7, in <module>
    time.sleep(1)
KeyboardInterrupt
pi@raspberrypi:~ $ more loop.log
I am still running :)
I am still running :)
I am still running :)
I am still running :)
I am still running :)
I am running again :)
I am running again :)
I am running again :)
I am running again :)
I am running again :)
pi@raspberrypi:~ $
```

Figure 6-29. *Using >> to append output to an existing file*

You can see that the output from the first script was added to by the output from the second script. However, the traceback error I caused when I exited the script with Ctrl-C did not get sent to the file. In order to capture errors as well as normal output, you will need to use the &> symbols to overwrite or &>> symbols to append to a file and include error messages (see Figure 6-30):

```
python loop.py &>> loop.log
```

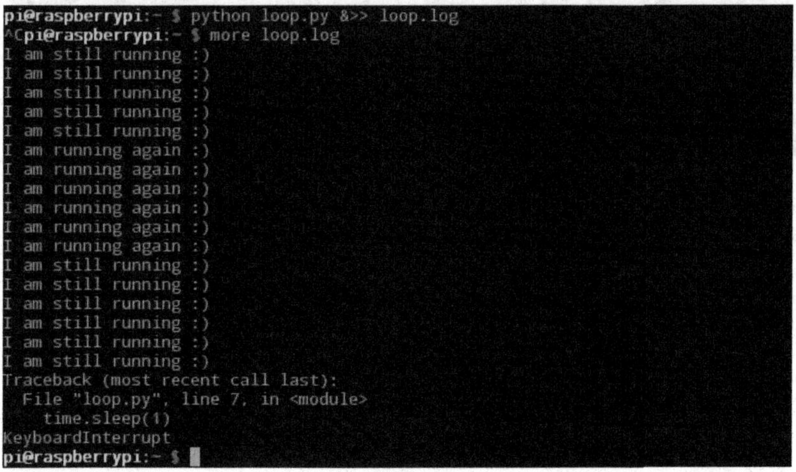

```
pi@raspberrypi:~ $ python loop.py &>> loop.log
^Cpi@raspberrypi:~ $ more loop.log
I am still running :)
I am still running :)
I am still running :)
I am still running :)
I am still running :)
I am running again :)
I am running again :)
I am running again :)
I am running again :)
I am running again :)
I am still running :)
I am still running :)
I am still running :)
I am still running :)
I am still running :)
Traceback (most recent call last):
  File "loop.py", line 7, in <module>
    time.sleep(1)
KeyboardInterrupt
pi@raspberrypi:~ $
```

Figure 6-30. *Using &>> to append standard output and errors to an existing file*

Searching the Output of a Command: grep

As you saw earlier, you can use the grep command to find a given search string in the output of the ps command. You can also use grep with almost any other command to search through the output it provides. The origins of the grep command are a bit more esoteric than other Linux commands. grep stands for *globally search a regular expression and print*. To use grep, type your command followed by a | symbol, then follow that with grep and with your search term. For example, if you have a logfile, you could search through it by using the more command to print the contents of the file, combined with the grep command to display only the lines that contain your search term (see Figure 6-31).

```
more loop.log | grep again
```

```
pi@raspberrypi:~ $ more loop.log | grep again
I am running again :)
I am running again :)
I am running again :)
I am running again :)
I am running again :)
I am running again :)
pi@raspberrypi:~ $
```

Figure 6-31. *Using grep to only print lines that have "again" in them*

There are many useful options you can use with `grep` as well. Here are a few of my favorites:

`-e`

Search for multiple terms at the same time.

`-i`

Run a case-insensitive search.

`-c`

Count how many lines contain the search term.

When you combine these options, `grep` becomes a very powerful tool (see Figure 6-32). There are even more useful options, which you can find in the manpage.

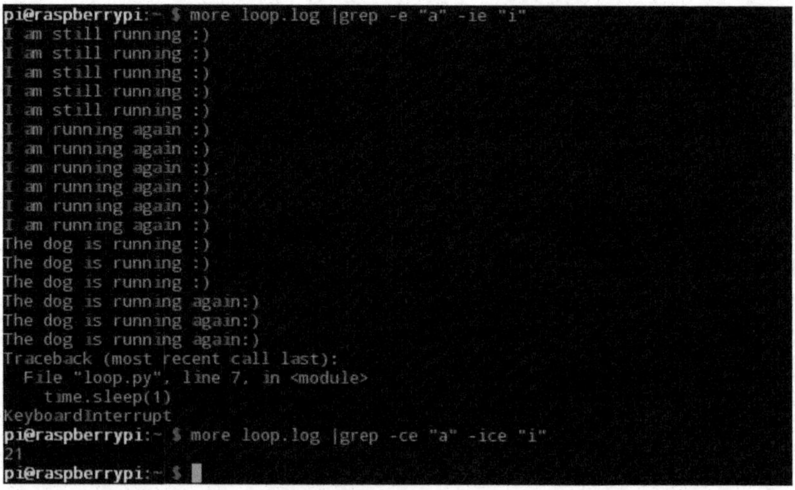

```
pi@raspberrypi:~ $ more loop.log |grep -e "a" -ie "i"
I am still running :)
I am still running :)
I am still running :)
I am still running :)
I am still running :)
I am running again :)
I am running again :)
I am running again :)
I am running again :)
I am running again :)
The dog is running :)
The dog is running :)
The dog is running :)
The dog is running again:)
The dog is running again:)
The dog is running again:)
Traceback (most recent call last):
  File "loop.py", line 7, in <module>
    time.sleep(1)
KeyboardInterrupt
pi@raspberrypi:~ $ more loop.log |grep -ce "a" -ice "i"
21
pi@raspberrypi:~ $
```

Figure 6-32. *Using grep to search for multiple patterns*

In Figure 6-32, I had **grep** search and print all lines in a logfile that contained either a lowercase *a* OR an upper- or lowercase *i*. I then did the same search, but instead of printing the matching lines, I printed the total number of lines that matched that search. You can also combine multiple **grep** statements to achieve an AND operation (see Figure 6-33).

```
pi@raspberrypi:~ $ more loop.log | grep dog | grep again
The dog is running again:)
The dog is running again:)
The dog is running again:)
pi@raspberrypi:~ $
```

Figure 6-33. *Using grep to search for multiple patterns (continued)*

Monitoring a Log File: tail

Sometimes it can be helpful to search the last few lines of a log-file to see what happened just before a script or program crashed or caused an error. You can do this easily on the command line with the **tail** utility. As the name suggests, **tail** without any options prints out the last 10 lines of a file (see Figure 6-34).

```
pi@raspberrypi:~ $ tail loop.log
The dog is running :)
The dog is running :)
The dog is running :)
The dog is running again:)
The dog is running again:)
The dog is running again:)
Traceback (most recent call last):
  File "loop.py", line 7, in <module>
    time.sleep(1)
KeyboardInterrupt
pi@raspberrypi:~ $
```

Figure 6-34. *Using tail to print the last few lines of a file*

There are two very useful options with **tail**. One is the -n num ber option, which lets you print any number of lines instead of just 10. The other is the -f option, which will print the last 10 lines but also keep adding lines as they are written to the file. This gives you a way to monitor a logfile as your script or program is running so you can see the information while it is happening.

Adding a User: adduser, addgroup

At some point, you may want to add another user to your Raspberry Pi. For example, you might want to give someone else access to the system without sharing the "pi" user's password, files, and settings. Or you might want to give your project its own identity on the computer. Linux was built as a multiuser operating system, so adding a new user is a very straightforward process. Simply run the adduser utility using sudo with the command:

    sudo adduser *username*

Then follow the prompts (see Figure 6-35).

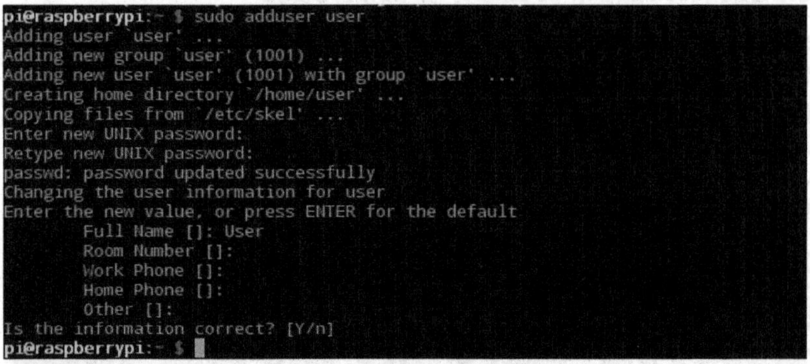

```
pi@raspberrypi:~ $ sudo adduser user
Adding user `user' ...
Adding new group `user' (1001) ...
Adding new user `user' (1001) with group `user' ...
Creating home directory `/home/user' ...
Copying files from `/etc/skel' ...
Enter new UNIX password:
Retype new UNIX password:
passwd: password updated successfully
Changing the user information for user
Enter the new value, or press ENTER for the default
        Full Name []: User
        Room Number []:
        Work Phone []:
        Home Phone []:
        Other []:
Is the information correct? [Y/n]
pi@raspberrypi:~ $
```

Figure 6-35. *Using adduser to add a new user in Linux*

When finished, you will have a new user account with its own home directory in */home* and a corresponding group with the same name. Likewise, if you only want to add a new group to the system, you can do that with the addgroup command (see Figure 6-36).

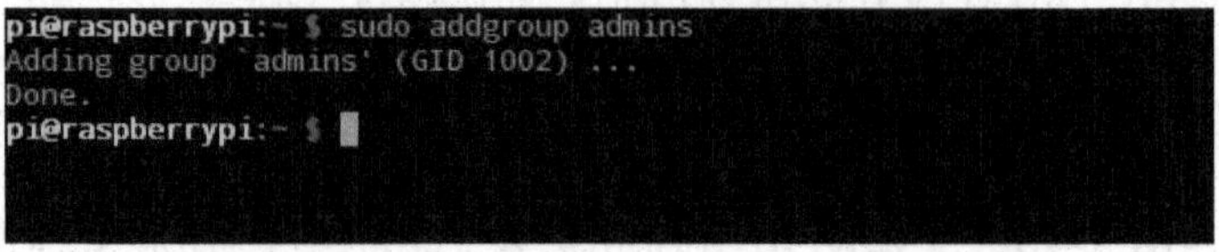

```
pi@raspberrypi:~ $ sudo addgroup admins
Adding group `admins' (GID 1002) ...
Done.
pi@raspberrypi:~ $
```

Figure 6-36. *Using addgroup to add a new group in Linux*

You can then add users to this new group by using the **adduser** command again. This time, follow the command with the username and then the group name (see Figure 6-37):

```
sudo adduser username groupname
```

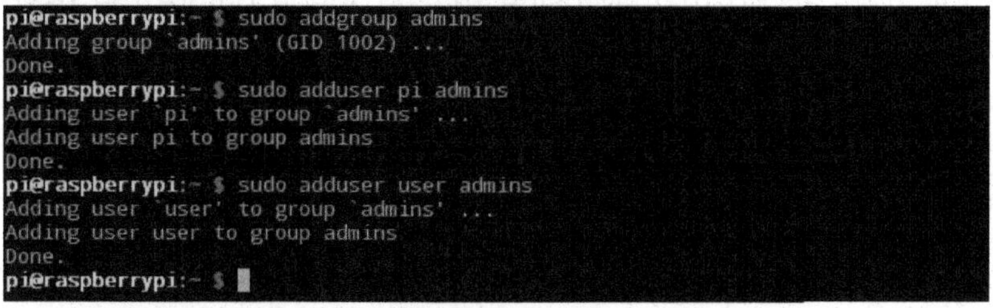

```
pi@raspberrypi:~ $ sudo addgroup admins
Adding group 'admins' (GID 1002) ...
Done.
pi@raspberrypi:~ $ sudo adduser pi admins
Adding user 'pi' to group 'admins' ...
Adding user pi to group admins
Done.
pi@raspberrypi:~ $ sudo adduser user admins
Adding user 'user' to group 'admins' ...
Adding user user to group admins
Done.
pi@raspberrypi:~ $ 
```

Figure 6-37. *Using adduser to add a user to a group*

Changing File Ownership and Permissions: chown, chmod

In Chapter 2, I explained how permissions work in Linux. Now let's take a look at how to change them. You might need to do this if you're getting an error like "Permission denied" when you try to run a script or command that is trying to access a file owned by another user. Some programs want to run as a separate user for security reasons, and you might need to change ownership of files so the program can access them. Also, if you're creating your own programs or scripts, you'll need to give them execute permission before they can be run.

Warning

Changing ownership or permissions for a system file that is normally only accessed by the "root" user could compromise the security of your system or lead to instability. You should use **sudo** to run those commands instead.

To change the ownership of a file (for a user and/or group), use the command **chown**. If you don't already have write permissions for the file, you will need to use **sudo** to change ownership because "root" can always perform these actions. You can change user and group ownership at the same time like this:

```
sudo chown <user>:<group> filename
```

If you only want to change the user-level ownership for a file, simply leave off the colon and the group name. It also can be convenient to change ownership for all the files in a given directory. You can do that by using the -R option before the user name:

```
sudo chown -R user directory
```

Keep in mind that, in Linux, each user also has a group automatically created for them with the same name as their username. This can be a bit confusing, but it does work nicely when you need to assign ownership to multiple users (see Figure 6-38).

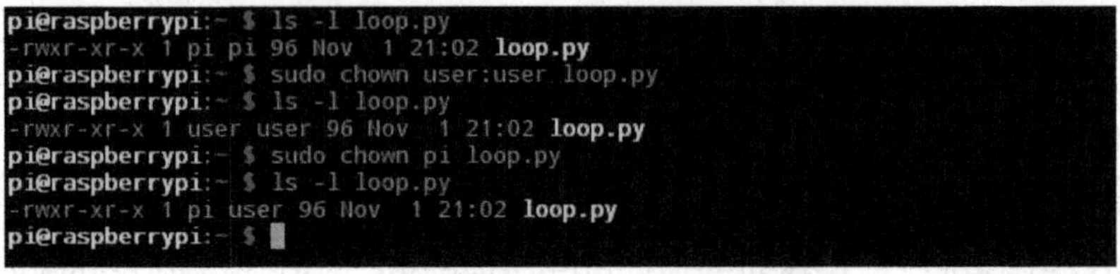

```
pi@raspberrypi:~ $ ls -l loop.py
-rwxr-xr-x 1 pi pi 96 Nov  1 21:02 loop.py
pi@raspberrypi:~ $ sudo chown user:user loop.py
pi@raspberrypi:~ $ ls -l loop.py
-rwxr-xr-x 1 user user 96 Nov  1 21:02 loop.py
pi@raspberrypi:~ $ sudo chown pi loop.py
pi@raspberrypi:~ $ ls -l loop.py
-rwxr-xr-x 1 pi user 96 Nov  1 21:02 loop.py
pi@raspberrypi:~ $ 
```

Figure 6-38. *Examples of using the chown command*

As you can see in Figure 6-38, I first changed user and group ownership for the file *loop.py* to the "user" user and the "user" group. Then I realized I wanted the "pi" user to keep ownership of the file, so I changed the user-level ownership to the "pi" user. Now, "pi" and "user" can both read and execute the *loop.py* file, but only "pi" can write to the file.

When you create a new file, it will be assigned the permissions 644 or rw-r--r--. This means that the owner can read and write to it, and everyone else can just read it. Notice that, by default, no one can execute the file. This is a problem if this is a script that you want to run as part of your project. So you will need to change the permissions by using chmod, which stands for *change file mode*. Similar to chown, you may need to use sudo to change a file's permissions.

You can use chmod to change permissions in two ways. One way is by specifying the numeric representation of the permissions you want to assign:

```
sudo chmod XXX filename
```

Here, *XXX* is the numeric permissions (i.e., 644). You can also add/remove a permission to all levels of ownership at the same time by using the + and - signs followed by the letters x, r, and/or w. So to add execute permissions for all users for a given file, you can type:

```
sudo chmod +x filename
```

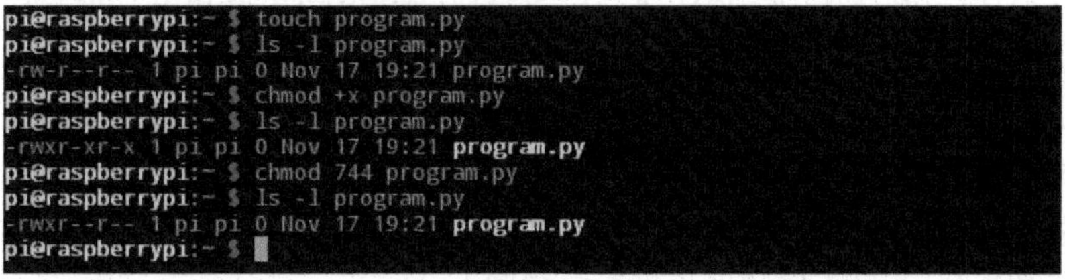

```
pi@raspberrypi:~ $ touch program.py
pi@raspberrypi:~ $ ls -l program.py
-rw-r--r-- 1 pi pi 0 Nov 17 19:21 program.py
pi@raspberrypi:~ $ chmod +x program.py
pi@raspberrypi:~ $ ls -l program.py
-rwxr-xr-x 1 pi pi 0 Nov 17 19:21 program.py
pi@raspberrypi:~ $ chmod 744 program.py
pi@raspberrypi:~ $ ls -l program.py
-rwxr--r-- 1 pi pi 0 Nov 17 19:21 program.py
pi@raspberrypi:~ $
```

Figure 6-39. *Using chmod to change file permissions*

As with chown, you can also change the permissions for entire directories by using the -r option. Be careful, though, since giving the wrong permissions to a file can lead to big security problems for your system.

Try It for Yourself

Create a new file and practice changing ownership and permissions.

Create a new file with touch:

```
touch program.py
```

Use ls to show the permissions and ownership:

```
ls -l program.py
```

Give the "root" group ownership of the file:

```
sudo chown pi:root program.py
```

Give all users execute permissions:

```
sudo chmod +x program.py
```

Use ls to verify your changes:

```
ls -l program.py
```

Running More Than One Command at the Same Time: &&, ||

Sometimes when you have a long-running program on the command line, it can feel like you're babysitting it. You are just staring at the screen waiting for the program or command to finish so you can run the next one based on whether the first one ran successfully or not. In these cases, it can be helpful to run both commands at the same time, so you can walk away and get a cup of coffee or get back to writing your book. The Linux shell has two built-in operators to help you do this. The first is represented by && and essentially means a logical AND. The other is represented by || and is like a logical OR.

The way this works on the command line is that if I have two commands separated by &&, the shell will run the command on the left side first to see if it ran successfully or not. If it did, the shell will run the command on the right side. If it didn't run successfully, the shell will not run the command on the right side. Just the opposite will happen if I separate two commands with ||. In this case, the command on the right will only run if the command on the left fails for some reason. You can also chain these together to get actions based on the results of a previous command (see Figure 6-40).

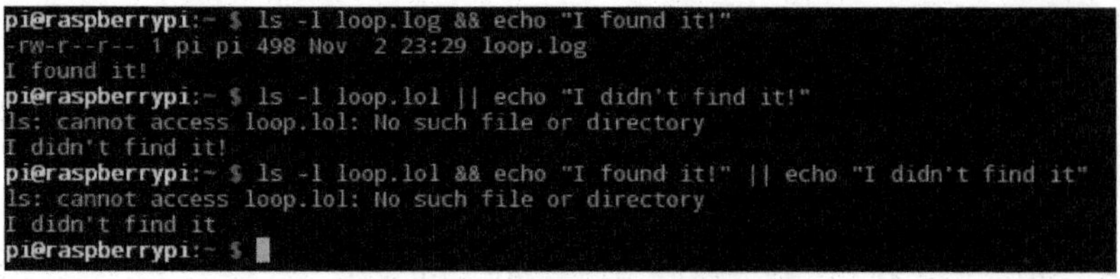

```
pi@raspberrypi:~ $ ls -l loop.log && echo "I found it!"
-rw-r--r-- 1 pi pi 498 Nov  2 23:29 loop.log
I found it!
pi@raspberrypi:~ $ ls -l loop.lol || echo "I didn't find it!"
ls: cannot access loop.lol: No such file or directory
I didn't find it!
pi@raspberrypi:~ $ ls -l loop.lol && echo "I found it!" || echo "I didn't find it"
ls: cannot access loop.lol: No such file or directory
I didn't find it
pi@raspberrypi:~ $ 
```

Figure 6-40. *Using && to run multiple commands sequentially*

You will see these operators used in startup scripts and other shell scripts, so it's good to know what they do even if you don't use them very often. An example of where a typical Maker might use this is when they want to update the software on their Raspberry Pi. As you know from Chapter 4, when you update your software, you should always run two commands (`sudo apt-get update` and `sudo apt-get upgrade`). Using these operators, we can chain these commands together to save some time like this:

```
sudo apt-get update && sudo apt-get -y upgrade
```

If you simply want to run two or more commands consecutively and you don't care about the outcome of the individual commands, you can separate each command with a semicolon (;). However, this is not recommended, as it can lead to all sorts of problems—most of the time when a command fails, it's a good idea to stop and figure out what went wrong.

Opening Another Console Session

Whether you're using the desktop or the command line, occasionally a program will misbehave and lock up, preventing you from using the keyboard and mouse in your current session. At that point, it can be difficult to determine whether your whole system is locked up or just the session you happen to be using at that moment. Instead of pulling out the power cord (which can potentially corrupt important system files), you can use a keyboard shortcut to switch to a different session and troubleshoot the problem from there.

When most distributions of Linux boot up, they actually start multiple virtual console sessions in the background. These are referred to in Linux as TTY1, TTY2, and so on. To display a different console session, press Ctrl-Alt-Func key on your keyboard (the Func key represents the F1 through F7 keys). When you do this, Linux will switch you to the corresponding console session TTY1 through TTY7. If you're running the desktop, it will be running in TTY7. If you are running without the desktop, you will be using TTY1. TTY2 through TTY6 are used for additional command-line console sessions.

So if you are on the desktop and it's locked up, press Ctrl-Alt-F1 to switch to the TTY1 console session. To go back to the desktop, press Ctrl-Alt-F7. Likewise, if you boot to the command prompt, then you're already using TTY1, so you can switch to another console session by pressing Ctrl-Alt-F2.

Direct Connect Only

In order for this to work, you need to be connected directly to the system. These commands won't work if you're connected remotely via SSH or VNC. They also won't work if your system has completely crashed and is blocking all keyboard input.

Dealing with Long Commands

As you can tell by now, some commands and programs have many, many options. Though this can be very powerful, it can also make for some very long commands that you have to type at the command prompt. Occasionally, these commands will wrap around your terminal window and make it hard to tell if you have a typo.

The Linux shell has a way to help you deal with this problem. By typing the escape character and pressing the Enter key, you can space out your command so that it doesn't wrap around the screen or just to make it easier to read. The escape character is the backslash (\) character on your keyboard. Just type it anywhere you want to break a line and press the Enter key. Then just keep typing until you're done with your command. You can do this as many times as you need to in order to keep your command neat and tidy (see Figure 6-41).

```
pi@raspberrypi:~ $ echo "Now is the time for all good men to come to the aid of
their country"
Now is the time for all good men to come to the aid of their country
pi@raspberrypi:~ $ echo "Now is the time \
> for all good men \
> to come to the aid \
> of their country"
Now is the time for all good men to come to the aid of their country
pi@raspberrypi:~ $ 
```

Figure 6-41. *Using the \ key to allow your commands to wrap to the next line*

My \ and Enter Keys Are Right Next to Each Other!

If you hit the escape character by accident and press Enter (like I do sometimes) but are at the end of the command, you can tell the shell the command is done by typing a semicolon (;) and pressing Enter again.

The escape character was created so that you could tell the shell to *escape* its interpretation of what you were typing and treat the next character in a different or more literal way. For example, you would typically use the double-quote (") character to enclose the text you want to print to the screen when using the echo command. However, if you want to actually print a double-quote character, you need to use the escape character first (see Figure 6-42).

```
pi@raspberrypi:~ $ echo ""Now" is the time"
Now is the time
pi@raspberrypi:~ $ echo "\"Now\" is the time"
"Now" is the time
pi@raspberrypi:~ $
```

Figure 6-42. *Using the \ key as an escape character*

There are lots of other great examples of how to use the escape character on the internet. Keep in mind that the escape character might behave differently depending on the environment you are using it in. So something that works in the Linux shell might work differently in a programming language like Python or Java.

Scheduling Jobs: cron

Running scripts from the command line is all well and good, but for some projects you will need to run a script at a set interval. Common uses for this would be to back up your project or files on a regular basis for safekeeping. Or perhaps you want to run a script every 10 minutes that gets a reading from a temperature sensor. For running scripts based on time, you can use a Linux utility called cron.

cron traces its roots back to the earliest days of Unix and is derived from the Greek word for time, *chronos*. cron runs in the background on Linux and is constantly keeping track of whether it's time to run a given script or command. Each user can configure cron individually by using a special text file called a *crontab*. This file cannot be edited with a normal text editor. Instead, a user edits their crontab by typing:

```
crontab -e
```

If this is the first time you've edited your crontab, the system will ask which editor you'd like to use (see Figure 6-43).

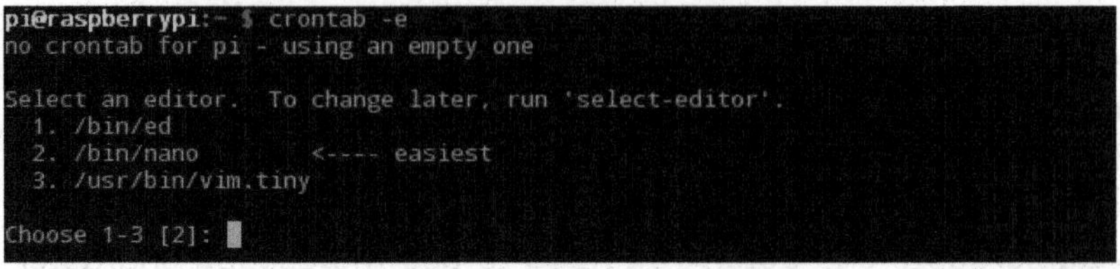

```
pi@raspberrypi:~ $ crontab -e
no crontab for pi - using an empty one

Select an editor.  To change later, run 'select-editor'.
  1. /bin/ed
  2. /bin/nano         <---- easiest
  3. /usr/bin/vim.tiny

Choose 1-3 [2]: █
```

Figure 6-43. *Choosing an editor for a crontab*

I will be using nano in this description of how a crontab works. Once you choose your editor, your crontab file will be loaded. The default file has a bit of text that's commented out as well as an example, but it can be a bit confusing to figure out what is going on. In a single line of text, cron looks for the minute, hour, day, month, and day of the week when the script or command should run. It's actually quite a flexible system once you know how to configure it. I will break down how this single line is formatted to schedule a task in Figure 6-44.

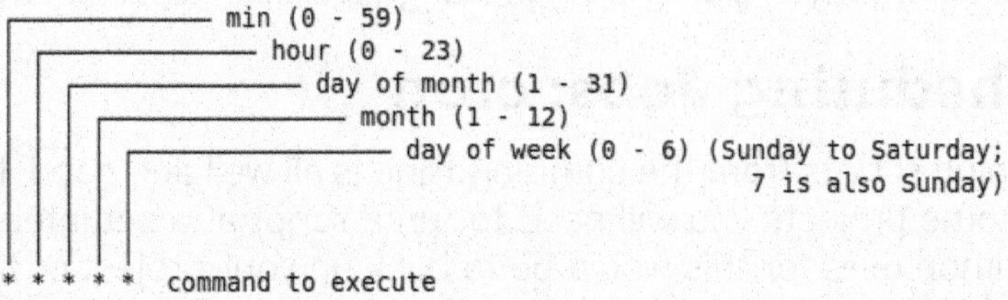

```
                  ┌──────────── min (0 - 59)
                  │ ┌────────── hour (0 - 23)
                  │ │ ┌──────── day of month (1 - 31)
                  │ │ │ ┌────── month (1 - 12)
                  │ │ │ │ ┌──── day of week (0 - 6) (Sunday to Saturday;
                  │ │ │ │ │                          7 is also Sunday)
                  │ │ │ │ │
                  * * * * *  command to execute
```

Figure 6-44. *Breakdown of the format of lines in the crontab file*

If I wanted to run the script *hello.sh* every Sunday in January at exactly 11:30 p.m., I would add this single line in my crontab (see Figure 6-45):

```
30 23 * 1 0 /home/pi/hello.sh
```

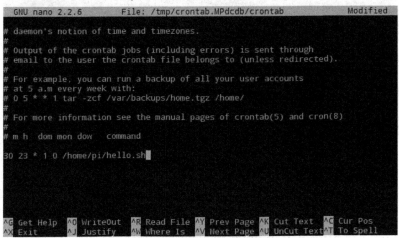

Figure 6-45. *Using cron to schedule a job*

Notice that I used an asterisk in the day-of-the-month position. An asterisk means "any." In this case, I didn't care which day of the month my script ran on. If I wanted to limit my script further to only run on Sundays that fall on the fifth day of the month, I would have put a 5 instead of the asterisk. You can also divide a particular position into increments by using the / character, or specify multiple values for the same position by using a comma. It is always a good idea to use the full path to your script. Here are some other examples of how to configure a line in your crontab (see Figure 6-46).

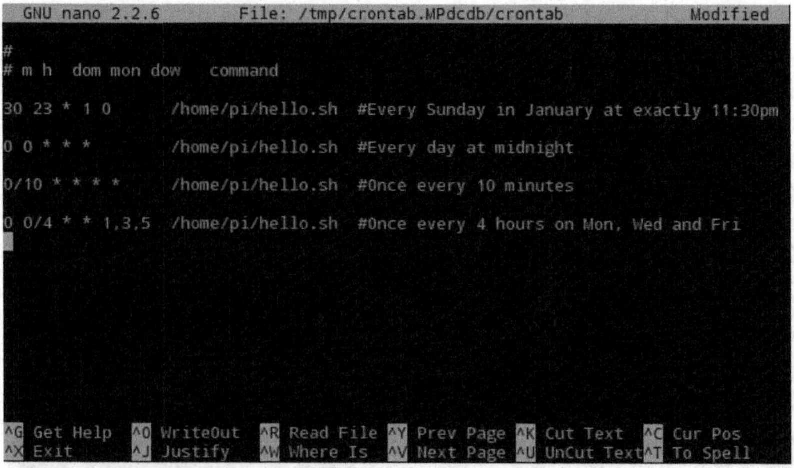

```
  GNU nano 2.2.6          File: /tmp/crontab.MPdcdb/crontab                    Modified

#
# m h  dom mon dow    command

30 23 * 1 0        /home/pi/hello.sh  #Every Sunday in January at exactly 11:30pm

0 0 * * *          /home/pi/hello.sh  #Every day at midnight

0/10 * * * *       /home/pi/hello.sh  #Once every 10 minutes

0 0/4 * * 1,3,5  /home/pi/hello.sh  #Once every 4 hours on Mon, Wed and Fri




^G Get Help  ^O WriteOut  ^R Read File ^Y Prev Page ^K Cut Text  ^C Cur Pos
^X Exit      ^J Justify   ^W Where Is  ^V Next Page ^U UnCut Text^T To Spell
```

Figure 6-46. *More examples of using cron*

 Don't Forget to Save

After you're done making changes, be sure to save your file. This will automatically update `cron` so it knows to check this file and run any scheduled jobs that are listed.

Why This Matters for Makers

As you build projects with Linux, you will eventually want to know how to monitor the performance of your system, add users and groups, change the permissions and ownership of files, and schedule jobs to run automatically. You might use some of them multiple times on every project, whereas others you might use rarely. In any case, knowing how to complete these tasks will help you solve problems as they come up. It will also speed up the time it takes to complete your project so you can enjoy your creation instead of troubleshooting it.

7/Controlling the Physical World

Most Makers will want to build a project that can manipulate and interact with the physical world by controlling sensors, motors, components, and devices. Controlling the devices and modules is mostly accomplished with programming, but there are some prerequisites to fulfill in Linux before you can jump into controlling them. In this chapter, I will explain how to control the general-purpose input/output (GPIO) pins, the inter-integrated circuit (I²C) protocol, and the serial peripheral interface (SPI) protocol and even how to interact with an Arduino. Though I won't be able to go into details about how the programming works, I will provide some examples in Python that illustrate the fundamentals of using these interfaces.

GPIO

One of the ways to control an external device is by using the GPIO pins that are built into the Raspberry Pi and many other SBCs and microcontrollers. A 40-pin header on one side of the Raspberry Pi provides an easy way to access the GPIO pins. However, of those 40 pins, only 26 are general-purpose input/output pins. The rest are voltage pins, ground pins, and pins that are only used by add-on boards. In addition, you can configure many of the pins to allow alternative functions like I²C and SPI (more on this in "I²C and SPI" on page 177) instead of GPIO. The physical pin numbers are not the same as the GPIO numbers, so it's important to have a reference when connecting devices to your Raspberry Pi (see Figure 7-1).

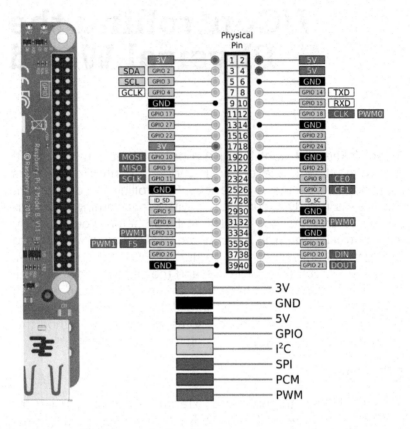

Figure 7-1. *Raspberry Pi B+ pinout*

As you can see in Figure 7-1, GPIO 2 is actually located on physical pin 3. If you tried to connect a device to physical pin 2, you would be connecting it to five volts of electricity, which might damage your device. You can also see where the alternative functions are located on the 40-pin header. So if you enable I²C functionality, you would use physical pins 3 and 5 to connect to your device, but you would lose GPIO 2 and 3 in the process.

There are several programming modules or libraries you can use to control the GPIO pins on a Raspberry Pi. One of the more popular is the Python module called *RPi.GPIO*. It is a nice, simple library used in many programming examples on the internet. However, I prefer another module called pigpio because it runs as a service on your system, and can be called from Python or C, or even from another Raspberry Pi across a network.

If you installed pigpio using `apt-get` in Chapter 4, you can skip this step. To install pigpio, you must download the latest version from the internet by using `wget`. `wget` stands for "web get" and can be used to download files from the internet on the command line as long as you know the complete URL for the file:

```
wget abyz.co.uk/rpi/pigpio/pigpio.zip
```

Now uncompress the file with `unzip`:

```
unzip pigpio.zip
```

Change directory into the folder you just unzipped:

```
cd PIGPIO
```

Then compile the software and install it:

```
make -j4
sudo make install
```

If you've never compiled software, you may be surprised at the strange output that's printed on the screen (see Figure 7-2). However, this is the normal output of many commands run in sequential order to build software.

```
strip --strip-unneeded libpigpiod_if.so
gcc -shared -o libpigpiod_if2.so pigpiod_if2.o command.o
size       libpigpiod_if.so
   text    data     bss     dec     hex filename
  59203    4292   49244  112739   1b863 libpigpiod_if.so
gcc -o pig2vcd pig2vcd.o
strip pig2vcd
strip --strip-unneeded libpigpiod_if2.so
gcc -o pigs pigs.o command.o
size       libpigpiod_if2.so
   text    data     bss     dec     hex filename
  69710    4292    1984   75986   128d2 libpigpiod_if2.so
gcc -o x_pigpiod_if x_pigpiod_if.o -L. -lpigpiod_if -pthread -lrt
strip pigs
gcc -o x_pigpiod_if2 x_pigpiod_if2.o -L. -lpigpiod_if2 -pthread -lrt
gcc -shared -o libpigpio.so pigpio.o command.o
strip --strip-unneeded libpigpio.so
size       libpigpio.so
   text    data     bss     dec     hex filename
 247360    5324  596672  849356   cf5cc libpigpio.so
gcc -o x_pigpio x_pigpio.o -L. -lpigpio -pthread -lrt
gcc -o pigpiod pigpiod.o -L. -lpigpio -pthread -lrt
strip pigpiod
pi@raspberrypi:~/PIGPIO $ sudo make install█
```

Figure 7-2. *Using the make command to build the pigpio program*

Now that pigpio is installed, you can run the service in the background like this:

```
sudo pigpiod &
```

If you want to run it every time your Raspberry Pi boots up, you can add the previously given line to your *rc.local* file as shown in Chapter 6. However, there is no need to use **sudo** if you are running it from *rc.local*, because *rc.local* runs as "root" already.

There are many coding examples available on the pigpio website (*http://bit.ly/2nAbESb*) for C, C++, and Python. One useful example is a small Python script that tells you the status of each GPIO pin. This will also ensure that the pigpiod service is running correctly. To use this script, open a new file with **nano**:

```
sudo nano gpio_status.py
```

Then type or copy the following code:

```
#!/usr/bin/python

import time
import curses
import atexit
import pigpio

GPIOS=32
MODES=["INPUT", "OUTPUT", "ALT5", "ALT4", "ALT0", "ALT1",
"ALT2",
"ALT3"]

def cleanup():
    curses.nocbreak()
    curses.echo()
    curses.endwin()
    pi.stop()

pi = pigpio.pi()
stdscr = curses.initscr()
curses.noecho()
curses.cbreak()
atexit.register(cleanup)
cb = []

for g in range(GPIOS):
    cb.append(pi.callback(g, pigpio.EITHER_EDGE))
```

```
# disable gpio 28 as the PCM clock is swamping the system

cb[28].cancel()
stdscr.nodelay(1)
stdscr.addstr(0, 23, "Status of gpios 0-31", curses.A_REVERSE)

while True:
    for g in range(GPIOS):
        tally = cb[g].tally()
        mode = pi.get_mode(g)
        col = (g / 11) * 25
        row = (g % 11) + 2
        stdscr.addstr(row, col, "{:2}".format(g), curses.A_BOLD)
        stdscr.addstr("={} {:>6}: {:<10}".format(pi.read(g),
        MODES[mode], tally))
    stdscr.refresh()
    time.sleep(0.1)
    c = stdscr.getch()
    if c != curses.ERR:
        break
```

Save and close the file by pressing Ctrl-X, then Y, then Enter. Now give the file execute permissions, as you learned in Chapter 6:

```
chmod 755 gpio_status.py
```

Now you can run the command and check the output (see Figure 7-3):

```
./gpio_status.py
```

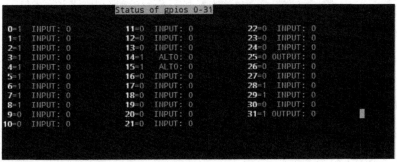

Figure 7-3. *The output of gpio_status.py*

You can see the status of all the GPIO pins and not just the 26 that are located on the 40-pin header. As you can see in Figure 7-3, most of the GPIO pins are registering as inputs. GPIO 14 and 15 are set to their `ALT0` function, which in this case is setting them to be transmit-and-receive serial communication pins (TXD and RXD).

Now let's put all this information to good use and make something happen. With a short Python script, you can make an LED blink or turn a relay on and off. Connect the positive pin of an LED to GPIO pin 18 (physical pin 12) and the negative pin to ground (see Figure 7-4).

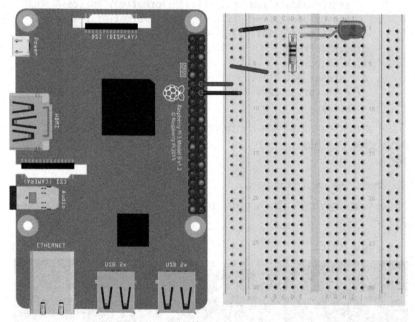

Figure 7-4. *Raspberry Pi with LED on GPIO 18*

Do I Need a Resistor?

The Raspberry Pi uses 3.3V on its GPIO pins. Some blue and white LEDs' forward voltage requirement is 3.3V, whereas other colors like yellow, red, and green run at lower voltages around 2V. If your LED requires less than 3.3V, you'll need to put a resistor between the positive pin of the LED and the GPIO pin on the Raspberry Pi. Always check the specifications of your LED to find the forward voltage. Although the LED will probably work without a resistor, it might not last very long. There are many good calculators online, like the one at *http://led.linear1.org/1led.wiz*, that can help you determine an adequate resistor value.

Now open a new file with **nano** for the Python script:

```
nano gpio_blink.py
```

Type or copy the following code:

```
#!/usr/bin/python

import pigpio
import time

pi = pigpio.pi()
# Set the GPIO mode as output
pi.set_mode(18, pigpio.OUTPUT)
# For relays, you want to set the initial mode to off
# Sometimes this means the pin is high or low depending on the
# relay
pi.write(18, 1)

# Now alternate on and off with half a second pause in between
while True:
    time.sleep(.5)
    pi.write(18, 0)
    time.sleep(.5)
    pi.write(18, 1)
```

Save and close the file by pressing Ctrl-X, then Y, then Enter. Now give the file execute permissions, as you learned in Chapter 6:

```
chmod 755 gpio_blink.py
```

Now you can run the command and check the output:

```
./gpio_blink.py
```

If everything is hooked up correctly, your LED should blink on for half a second and then off for half a second and then repeat.

You could also use this script to connect your Raspberry Pi to a 5V relay module that turns on and off. Simply connect your relay module input to the same GPIO pin and connect the 5V pin on the relay module to the 5V pin on the Raspberry Pi (see Figure 7-5).

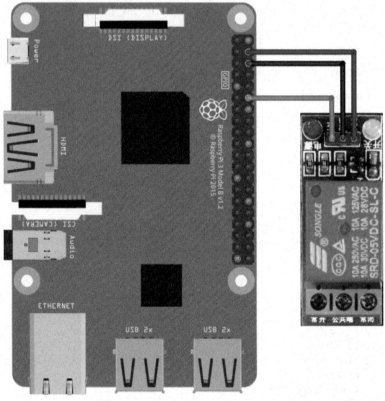

Figure 7-5. *Raspberry Pi with attached relay module*

A relay module, when completely connected, is a simple switch. A signal from the Raspberry Pi activates a small 5V switch, which in turn completes a circuit to control 120V mains power from a wall outlet, which turns on like a floor lamp or string of lights.

 CAUTION

Be extremely careful when using mains power! Be sure to turn off the power before you start working on the connections.

I²C and SPI

I²C and SPI are both serial communications protocols used to transmit data back and forth from one board to another. You can find all sorts of premade modules for your Raspberry Pi that use one of these protocols to communicate. I²C isn't as fast as SPI, but it has the benefit of requiring only two wires to connect multiple devices since each device has its own unique address. SPI, on the other hand, is much faster but requires four wires or more if you want to connect multiple SPI devices to the same board. In practice, I've never had to connect more than one device at a time, and the speed of communication isn't really an issue for most projects. So if I can find a module that supports I²C, I usually prefer it over SPI. However, it is good to know how to connect both types of devices.

In order to start using either of these protocols, you need to enable them on your Raspberry Pi. Remember how I mentioned alternative functions of the GPIO pins? Now we will change some of them to support I²C, SPI, or both at the same time. Note that once you do this, you won't be able to use those pins as normal GPIO until you change them back.

To enable I²C or SPI, start by running `raspi-config` in the console or terminal emulator (see Figure 7-6):

```
sudo raspi-config
```

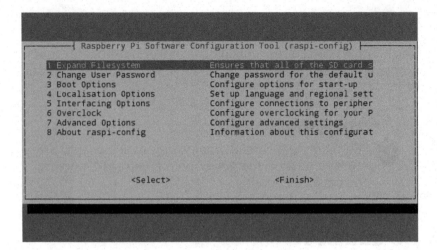

Figure 7-6. *The main raspi-config screen*

Use the arrow keys to move the cursor down to Interfacing Options and press the Enter key. Now move the cursor down to the SPI or I²C option, depending on which one you want to enable, and press Enter (see Figure 7-7).

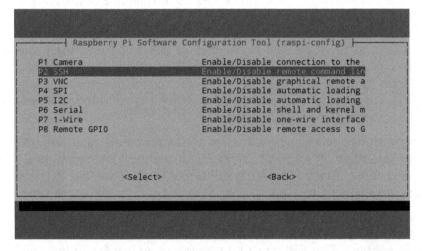

Figure 7-7. *The Interfacing Options menu in the raspi-config tool*

The `raspi-config` tool will ask if you want to enable the SPI or I²C interface. When you choose Yes, the `raspi-config` tool will make the necessary changes and let you know that the interface has been enabled. Press Enter to return to the main menu and use the arrow keys to select Finish to exit `raspi-config`.

If you're using I²C with an older version of the Raspbian distribution or a different Linux distribution, you might need to install some tools that may be required by the programming libraries you'll use to control your device. You can install these by running:

```
sudo apt-get update
sudo apt-get install i2c-tools
```

The last step is to add your user to a number of groups just to make sure you have the correct permissions to various files and software needed to communicate with your device:

```
sudo usermod -a -G i2c pi
sudo usermod -a -G spi pi
sudo usermod -a -G gpio pi
```

You need to reboot at this point to make sure the SPI or I²C interface is enabled and ready for use:

```
sudo shutdown -r now
```

Now that the interfaces are enabled, you can begin to use them.

 Know Your Device

Getting the Raspberry Pi ready to communicate to a device using I²C or SPI is one thing. Knowing *what* to say is quite another. Each device will require different instructions to perform the actions the device provides. Finding a programming library written with your device in mind will make this job much easier, especially for beginners.

Since there are so many different devices and modules you can use with the Raspberry Pi, there's no way for me to cover them all. So let's take a look at just one example to give you an idea of what is required and how using these protocols works. For this example, I will be connecting a small 128×64-pixel OLED display module to my Raspberry Pi using the I²C protocol. This display could be used in a small project to show minimal information like the weather or the status of a program you're running. I'll be using Python to display a simple message on the screen when my program is run.

This particular display uses a display driver called SSD1306, so I need to find a Python module that supports this particular driver. Luckily, there are two good Python modules you can use for these displays. One of them is from Adafruit (*http://bit.ly/2nkmlJ3*). It was written to go along with the SSD1306 display modules they sell. The other module (*http://bit.ly/2nhYfOe*) was written by Richard Hull. It supports a number of similar display drivers as well as the SSD1306. For this example, I will use the second one, as Hull has included a number of fun demo programs on GitHub (*http://bit.ly/2nBlVAl*) that show how to use the display.

First, install the supporting software and make sure it's up-to-date:

```
sudo apt-get update
sudo apt-get install python-dev python-pip libfreetype6-dev
libjpeg8-dev libsdl2-dev
sudo pip install --upgrade luma.oled
```

Now we can use the information on the Python module's website (*http://bit.ly/2nhYfOe*) to create a small test program. Open a new file with **nano** for the Python script:

```
nano ssd1306_example.py
```

Type or copy the following code:

```
#!/usr/bin/python

from luma.oled.serial import i2c
from luma.oled.device import ssd1306, ssd1331, sh1106
from luma.oled.render import canvas

# rev.1 users set port=0
# substitute spi(device=0, port=0) below if using that
# interface
serial = i2c(port=1, address=0x3C)

# substitute ssd1331(...) or sh1106(...) below if using
# that device

device = ssd1306(serial)
```

```
while True:
    with canvas(device) as draw:
        draw.rectangle(device.bounding_box, outline="white",
        fill="black")
        draw.text((30, 40), "Hello World", fill="white")
```

Save and close the file by pressing Ctrl-X, then Y, then Enter. Give the file execute permissions, as you learned in Chapter 6:

```
chmod 755 ssd1306_example.py
```

Now connect the SSD1306 display to the Raspberry Pi as shown in Figure 7-8.

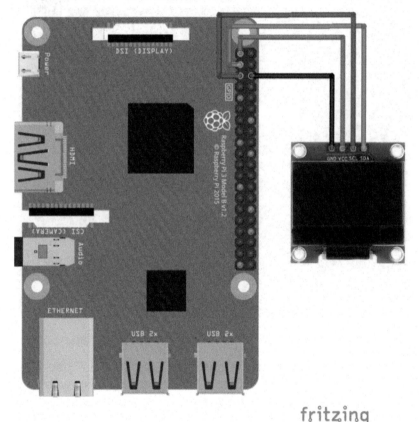

Figure 7-8. *Raspberry Pi with attached SSD1306 OLED display*

Now you can run the command and check the output:

```
./ssd1306_example.py
```

The OLED display should now show a rectangle with the words "Hello World" inside (see Figure 7-9).

Figure 7-9. *An SSD1306 I²C module running the example code*

I mentioned before that the author of this software package, Richard Hull, has some example scripts you can run with this type of display. If you're setting up a similar display and want to try them, you must first download the source code for the Python module. You can do this with wget:

```
wget https://github.com/rm-hull/luma.examples/archive/
master.zip
```

Then unzip the archive:

```
unzip master.zip
```

Change directory into the newly created *luma.examples-master/examples* directory:

```
cd luma.examples-master/examples
```

Then run one of the example programs to see what it looks like (see Figure 7-10):

```
./demo.py
```

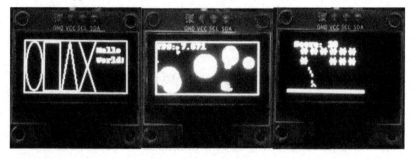

Figure 7-10. *Running the example scripts demo.py, bounce.py, and invaders.py on an SSD1306 display*

Try It for Yourself

Find a module that uses I2C or SPI and try getting it working with the Raspberry Pi. You could try a display, a temperature sensor, an accelerometer, or many others. Be sure to note if the device uses a particular driver or chipset and use that information to search the internet for a programming library or module that you can use to make your device work.

Talking to Arduino

Sometimes you need to connect a Raspberry Pi to an Arduino, either to improve reliability or processing speed. Or perhaps you are just more comfortable using an Arduino to control sensors, motors, and so on. You can still benefit from the flexibility of a Raspberry Pi and use an Arduino at the same time by setting up the Raspberry Pi to communicate with the Arduino using the I2C protocol.

Double-Check Your Connections

Arduino pin signals generally run at 5V, while the Raspberry Pi's run at 3.3V. If you're running the I²C bus with the Raspberry Pi as the master, as shown in the following example, everything should be fine. However, connecting a 5V signal to the wrong pin could damage your Raspberry Pi. Double-check your connections to make sure everything is wired up correctly. In general, if you want to connect 5V signals to 3.3V pins, you should use a logic-level voltage converter in between.

For this exercise, I will show you how to run the Raspberry Pi as the I²C master requesting information and an Arduino as the I²C slave that will be sending the information. Let's start by programming the Arduino. Using the Arduino IDE, create a new sketch using the following code and upload it into an Arduino:

```
#include <Wire.h>

void setup()
{
  Wire.begin(8);                   // join i2c bus with address #8
  Wire.onRequest(requestEvent);    // register event
}

char str[17];
int x = 0;

void requestEvent() {
  sprintf(str, "Message %7d\n", x);
  if (++x > 9999999) x=0;
  Wire.write(str);                 // sends 16 bytes
}

void loop() {
    delay(50);
}
```

This sketch will tell the Arduino to act as a slave on the I²C bus and respond to requests with the word "Message" along with a number that increments every time a request is made. This simulates data that would otherwise be coming from a sensor or other device connected to the Arduino.

Now open a file on the Raspberry Pi for the Python code that will make the requests to the Arduino:

```
nano i2c_master.py
```

Type or copy the following code:

```python
#!/usr/bin/python

import time
import pigpio

BUS=1
I2C_ADDR=8

pi = pigpio.pi()
# Open the connection to slave
h = pi.i2c_open(BUS, I2C_ADDR)

while True:
    # Make a generic request without registers
    (c, d) = pi.i2c_read_device(h,16)
    if c >= 0:
        print d
    else:
        print "No data ..."
    time.sleep(.5)

pi.i2c_close(h)
pi.stop()
```

Save and close the file by pressing Ctrl-X, then Y, then Enter. Now give the file execute permissions, as you learned in Chapter 6:

```
chmod 755 i2c_master.py
```

If you haven't done so already, start the pigpiod service like so:

```
sudo pigpiod &
```

Make the connections between your Raspberry Pi and Arduino using Figure 7-11.

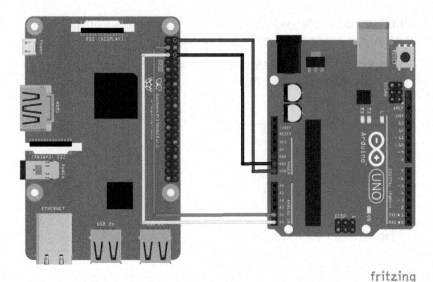

fritzing

Figure 7-11. *Connections for Raspberry Pi I²C master to Arduino slave*

Now you can run the command and check the output:

```
./i2c_master.py
```

The Python script uses the pigpio program we installed earlier to make continuous requests to the Arduino and print the response that it receives (see Figure 7-12). If this were actual data, you could then use this in the rest of your program. If the script doesn't receive a response, it prints "No data...".

```
pi@raspberrypi:~ $ sudo pigpiod &
[1] 1224
pi@raspberrypi:~ $ ./i2c_master.py
Message        0

Message        1

Message        2

Message        3

Message        4

Message        5

Message        6

Message        7

Message        8

Message        9

Message       10
```

Figure 7-12. *The Raspberry Pi receiving messages from an Arduino*

Why This Matters for Makers

Many projects that Makers build include interacting with devices in an attempt to bring the project to life. Knowing how to use the GPIO pins to control and communicate with other devices while navigating the Linux command line will help you finish your project more quickly and open up many new possibilities to be creative. Because protocols like I²C and SPI are standards, you can learn to communicate with thousands of different modules that provide functionality not natively available on the Raspberry Pi.

8/Using Multimedia

A great way to add life to a project is to add sound and video. Recently, I was working on a laser maze exhibit that some colleagues and I take to various Maker Faires and other events around the region. This massive 10×20-foot enclosure is full of laser beams, smoke, and buttons. This is all very cool in and of itself, but we found that by adding a start sound, stop sound, and a little spy-themed music, we could make the exhibit much more engaging.

No matter what you're making, from a simple doorbell to a complex video kiosk, you will need to know how to incorporate media into your project. There are multiple ways to interact with multimedia on Linux. In this chapter, I will discuss some of the pitfalls and things to watch out for as well as the most effective ways for Makers to use multimedia in their projects.

Choosing HDMI or Analog

The first thing you need to decide when using audio in your project is where you want the audio to go. By default, on a Raspberry Pi the audio will try to play through your HDMI connection. While this is nice if you're using an HDMI-capable display in your project, many times you won't have a monitor attached at all or your monitor might not have speakers. In this case, you will need to tell your Raspberry Pi to use the analog audio jack to output sound.

To do this, run the Raspberry Pi configuration script from the command line:

```
sudo raspi-config
```

This opens the Raspberry Pi configuration tool (see Figure 8-1).

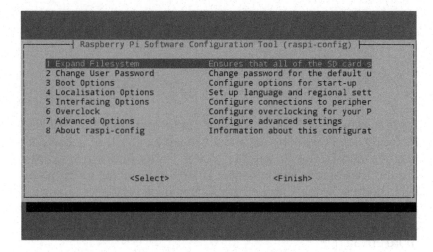

Figure 8-1. *The raspi-config main menu*

Use the arrow keys to move the cursor down to Advanced Options and press the Enter key. Now move the cursor down to Audio and press Enter again (see Figure 8-2).

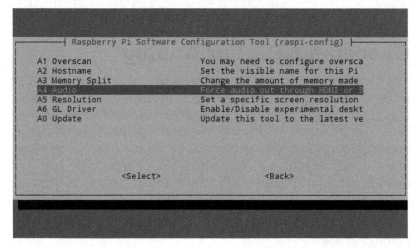

Figure 8-2. *The raspi-config Advanced Options menu*

Select the "Force 3.5mm ('headphone') jack" option and press Enter (see Figure 8-3).

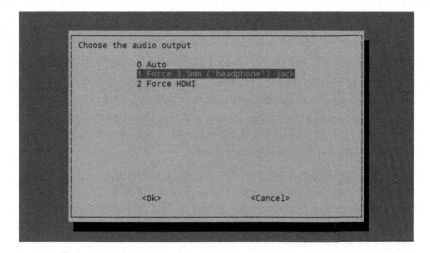

Figure 8-3. *The raspi-config Audio menu*

This will make the analog jack the default for sound that you play. You can still override this on an individual basis if you need to (see the next section). You'll need to reboot your system in order for the changes to take effect.

Playing Audio and Video Files

Linux has many utilities to play audio and video files. For Raspberry Pi, it's a good idea to use omxplayer, which comes installed with the Raspbian distribution. The first reason for this is that omxplayer has the ability to make use of the graphics processing unit (GPU), which will make playing HD-quality videos much less taxing on the system. Second, omxplayer has the ability to decode many digital formats like MP3 and MP4, which saves time by not having to deal with multiple utilities to play different file formats.

I often get asked whether people should install mplayer or vlc to play video files on the Raspberry Pi. The short answer is no. The reason is because, as of this writing, those utilities cannot use the GPU so they must decode the video file in software mode, which uses up a lot of CPU time and can cause other programs to become unresponsive.

To play a media file with `omxplayer`, use the following command:

```
omxplayer -o [local | hdmi] filename
```

Here, you would use `local` if you want the audio to play out of the analog audio port or `hdmi` to play it out of the HDMI-connected display. `omxplayer` doesn't always obey the default you chose in `raspi-config`, so it's a good idea to always specify this. Also, you could use this to override your default setting for audio output on a case-by-case basis.

After you run the command, the audio or video will display some encoding information about the file and it will play. When the file is done playing, you will be returned to the command prompt. If you want to stop playing the file before it's finished, press the Q key (see Figure 8-4).

```
pi@raspberrypi:~ $ omxplayer -o local tts.mp3
Audio codec mp3 channels 1 samplerate 24000 bitspersample 16
Subtitle count: 0, state: off, index: 1, delay: 0
have a nice day ;)
pi@raspberrypi:~ $ omxplayer -o local Nyan\ Cat.mp4
Video codec omx-h264 width 540 height 360 profile 578 fps 29.970030
Audio codec aac channels 2 samplerate 44100 bitspersample 16
Subtitle count: 0, state: off, index: 1, delay: 0
V:PortSettingsChanged: 540x360@29.97 interlace:0 deinterlace:0 anaglyph:0 par:1.
00 display:0 layer:0 alpha:255 aspectMode:0
Stopped at: 00:00:17
have a nice day ;)
pi@raspberrypi:~ $ █
```

Figure 8-4. *Using omxplayer to play audio and video files*

 No Video?

If you're connected via SSH or VNC, you won't see any video because the video is sent to the HDMI or analog video port.

Controlling the Volume

There are two ways to control the volume on a Raspberry Pi from the command line. To change the volume of the entire system, you can use the `alsamixer` utility. When you run this command, you will see a graphical representation of the volume level of the sound device that is built into the Raspberry Pi (see Figure 8-5).

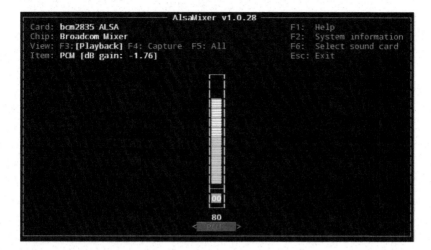

Figure 8-5. *Using alsamixer on the Raspberry Pi*

You can use the up and down arrows to raise or lower the volume or press the M key to mute the volume. When you are done, press the Esc key to exit.

Another way to control the volume is to use omxplayer to change the volume of a file as it's being played. This can be helpful if most of your media is playing at the right volume but you have one file that needs to be played a little louder or quieter. Just use omxplayer as before, but add the --vol option:

```
omxplayer -o [local | hdmi] --vol <millibels> filename
```

In this case, *millibels* represents a number between 500 (loud) and −4,000 (very quiet) with 0 being normal. If you use numbers outside of this range, your audio is likely to sound distorted.

Playing Media from a Script

Whether you like to program in Python, Perl, Java, Go, or Ruby, there are many ways to play media from a script. Most programming languages have a way to send commands to the underlying operating system. This may not always be the best choice, but for small projects running on an SBC, it can be the easiest way to get predictable results. Given what you've just learned, you can now use those commands to play media on your Raspberry Pi from a script. Here is an example in Python:

```python
#!/usr/bin/python

import os

os.system("omxplayer -o local filename.mp3")
```

In this case, the `os.system` method simply runs the command inside the quotation marks on the local operating system, which plays the MP3 file.

Why This Matters for Makers

As you build projects, it is important to think about the aesthetics of what you're building. Many times, that can include audio or video elements. Knowing how to use the command line to play audio and video can help bring your projects to life for those who are using them.

9/Accessing Cloud Services

In the age of the Internet of Things (IoT), interacting with cloud services or even creating your own cloud with Linux is becoming an important component of many projects. Perhaps you'd like to sync local files on your Raspberry Pi to a cloud storage service, or set up your own file storage service for your home that the whole family can use. Maybe you'd like to use a Raspberry Pi to turn on your sprinkler system from anywhere in the world, or send you a text message when the lights at home have been on too long. In this chapter, I'll get you started down the road of accessing and using cloud services with Linux.

Cloud Storage Services from the Command Line

If you've been writing your own programs for a while like I have, you've probably already started storing them both on your local machine and somewhere in the cloud for safekeeping. Sometimes I want to work on a program from home and then later put that program on my Raspberry Pi to test it out. If my program is in the cloud, I can access it using software that syncs that remote file with a local one.

The confusing wonder of it is that there are numerous services out there that store your files in the cloud. I use Google Drive, but perhaps you prefer Dropbox or some other service. As of this writing, Google Drive still hasn't released a native client for Linux. Dropbox has a GUI client, but there's no way to interact with it programmatically. This is a problem if you want to write a backup script using `cron` but can't interact with the service you want to use. So even though there are multiple software packages for all the various storage services that exist out there, I'm going to show you one tool you can use to interact with files on many different services.

rclone is a command-line program written in the Go language that can interact with a variety of cloud storage services to copy files or sync entire directories. It can also mount cloud locations locally and sync between two different cloud storage services (i.e., Google Drive to Dropbox). rclone currently supports the following services:

- Amazon Drive
- Amazon S3
- Backblaze B2
- Dropbox
- Google Cloud Storage
- Google Drive
- Hubic
- The local filesystem
- Microsoft One Drive
- Openstack Swift/Rackspace cloud files/Memset Memstore
- Yandex Disk

To install rclone, first download the latest version by using wget. As mentioned earlier, wget stands for "web get" and can be used to download files from the internet on the command line as long as you know the complete URL for the file:

```
wget http://downloads.rclone.org/rclone-current-linux-arm.zip
```

Uncompress the downloaded file with the unzip utility. This will extract the files inside a new directory in your current location:

```
unzip rclone-current-linux-arm.zip
```

Now change your location to the new directory that was just created. The version number might be different than the following one, so use ls or autocomplete to help you get the right name of the directory:

```
cd rclone-v1.34-linux-arm
```

rclone will need to be configured before you can begin using it. To start the configuration, run rclone with the config option. Figure 9-1 shows an example of what this looks like for Google

Drive; you can find individual setup guides for the other storage services on the rclone website (*http://rclone.org*):

```
./rclone config
```

```
pi@raspberrypi:~/rclone-v1.34-linux-arm $ ./rclone config
2016/12/08 18:26:51 Failed to load config file "/home/pi/.rclone.conf" - using d
efaults: open /home/pi/.rclone.conf: no such file or directory
No remotes found - make a new one
n) New remote
s) Set configuration password
q) Quit config
n/s/q> n
name> gdrive
Type of storage to configure.
Choose a number from below, or type in your own value
 1 / Amazon Drive
   \ "amazon cloud drive"
 2 / Amazon S3 (also Dreamhost, Ceph, Minio)
   \ "s3"
 3 / Backblaze B2
   \ "b2"
 4 / Dropbox
   \ "dropbox"
 5 / Encrypt/Decrypt a remote
   \ "crypt"
 6 / Google Cloud Storage (this is not Google Drive)
   \ "google cloud storage"
 7 / Google Drive
   \ "drive"
 8 / Hubic
   \ "hubic"
 9 / Local Disk
   \ "local"
10 / Microsoft OneDrive
   \ "onedrive"
11 / Openstack Swift (Rackspace Cloud Files, Memset Memstore, OVH)
   \ "swift"
12 / Yandex Disk
   \ "yandex"
Storage> 7
Google Application Client Id - leave blank normally.
client_id>
Google Application Client Secret - leave blank normally.
client_secret>
Remote config
Use auto config?
 * Say Y if not sure
 * Say N if you are working on a remote or headless machine or Y didn't work
y) Yes
n) No
```

Figure 9-1. *rclone configuration example for Google Drive*

When the configuration script starts, press the N key to set up a new remote service, then give it a descriptive name. Choose the service you want to set up. In this case, I chose Google Drive. At this point, the type of information needed will vary from service to service. You may need to supply an API key or username/ password in order to connect to the storage service you want to set up. For Google Drive, you can simply press Enter when the configuration script asks for client_id and client_secret.

When the script asks for "auto config," you can press the N key, but if you're on the desktop it will try to open a browser window to complete this step anyway. Next, you will be presented with a URL to put in a browser to get the authorization code you need in order to proceed (see Figure 9-2).

```
y/n> n
If your browser doesn't open automatically go to the following link: https://acc
ounts.google.com/o/oauth2/auth?client_id=              .apps.googleusercontent.com
&redirect_uri=urn%3A1etf%3Awg%3Aoauth%3A2.0%3Aoob&response_type=code&scope=https
%3A%2F%2Fwww.googleapis.com%2Fauth%2Fdrive&state=
3
Log in and authorize rclone for access
Enter verification code>
--------------------
[gdrive]
client_id =
client_secret =
token = {"access_token":"
             ","token_type":"Bearer","refresh_token"
             ","expiry":"2016-12-08T19:41:31.413064152Z"}
--------------------
y) Yes this is OK
e) Edit this remote
d) Delete this remote
y/e/d> y
Current remotes:

Name                    Type
====                    ====
gdrive                  drive

e) Edit existing remote
n) New remote
d) Delete remote
s) Set configuration password
q) Quit config
e/n/d/s/q>
```

Figure 9-2. *rclone configuration example for Google Drive (continued)*

This URL will take you to your Google Drive account where you will need to sign in and grant access to `rclone` for remote access. You will be given the verification code to enter on the next line. In Figure 9-2, I have hidden the sensitive parts of the configuration, but your screen should look similar. Confirm your settings by pressing the Y key and then quit the configuration script.

Now you can use `rclone` to transfer files to and from your storage service. `rclone` will compare the files and transfer only the ones that have changed. The syntax for `rclone` is similar to `cp`. Here are some useful commands to use with `rclone`.

List files:

```
rclone ls <remote name>:
rclone ls <remote name>:<directory>
```

List only directories:

```
rclone lsd <remote name>:
rclone lsd <remote name>:<directory>
```

Copy files from one location to another:

```
rclone copy <remote name>:<directory> <local directory>
rclone copy <local directory> <remote name>:<directory>
```

In this case, I want to transfer some sound files from Google Drive to my Raspberry Pi. So I create a subdirectory on my Raspberry Pi and copy the files to it (see Figure 9-3).

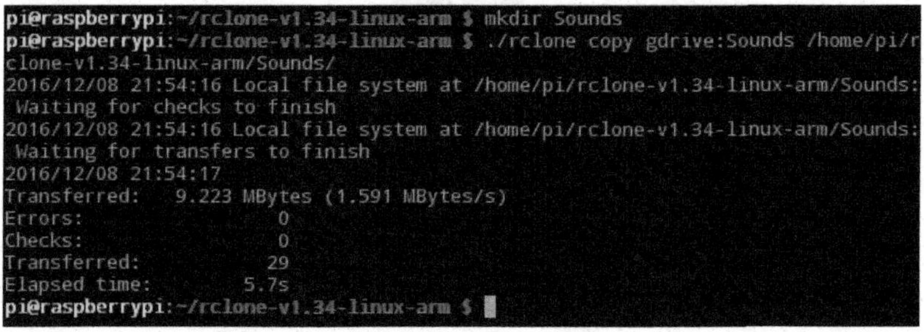

```
pi@raspberrypi:~/rclone-v1.34-linux-arm $ mkdir Sounds
pi@raspberrypi:~/rclone-v1.34-linux-arm $ ./rclone copy gdrive:Sounds /home/pi/r
clone-v1.34-linux-arm/Sounds/
2016/12/08 21:54:16 Local file system at /home/pi/rclone-v1.34-linux-arm/Sounds:
Waiting for checks to finish
2016/12/08 21:54:16 Local file system at /home/pi/rclone-v1.34-linux-arm/Sounds:
Waiting for transfers to finish
2016/12/08 21:54:17
Transferred:      9.223 MBytes (1.591 MBytes/s)
Errors:                0
Checks:                0
Transferred:          29
Elapsed time:       5.7s
pi@raspberrypi:~/rclone-v1.34-linux-arm $ █
```

Figure 9-3. *Using rclone to copy files*

IFTTT

IFTTT (aka If This, Then That) is a web-based service that allows you to connect IoT devices through the cloud and trigger certain activities based on given criteria. For example, when a temperature sensor connected to a Raspberry Pi reaches a certain level, you could send yourself an email letting you know the time and date that level was first reached. Or you could tell a digital assistant (Amazon Alexa, Google Home) to open the garage door, and IFTTT would send a signal to your Raspberry Pi to trigger a relay to open the door. It's possible to do these things without IFTTT, but using IFTTT makes it a lot easier since it already interacts with so many different services and devices.

To enable your system to work with IFTTT, you need to set up either incoming or outgoing communication, or both. To do so, you can use a simple command-line tool called `curl` to send commands to IFTTT. `curl` stands for *command-line URL* and is fine for testing purposes. For more advanced usage, you can run a Python-based web server that responds to incoming commands and sends outgoing messages to IFTTT.

Before you can use your Raspberry Pi with IFTTT, you need to set up an account and subscribe to the Maker service. There are many guides on the internet to help you with these tasks. After you subscribe to the Maker service, go to the settings for the service and make note of your URL (see Figure 9-4).

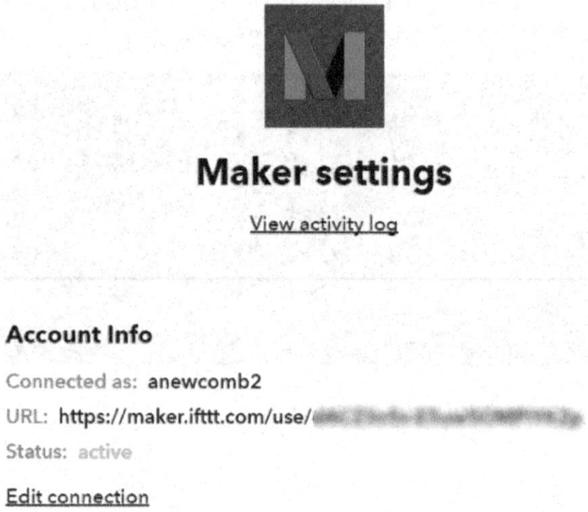

Maker settings

View activity log

Account Info

Connected as: **anewcomb2**

URL: **https://maker.ifttt.com/use/**

Status: active

Edit connection

Figure 9-4. *Maker service settings on IFTTT*

Now you need to create an applet that you can trigger from your Raspberry Pi. Click on your username and choose New Applet to start the Applet Maker. Click on the blue "this" and choose the Maker service (see Figure 9-5).

Applet Maker

if ➕ this then that

Want to build even richer Applets?

Figure 9-5. *The initial Applet Maker page*

After you select the Maker service, there will be only one trigger to choose (see Figure 9-6).

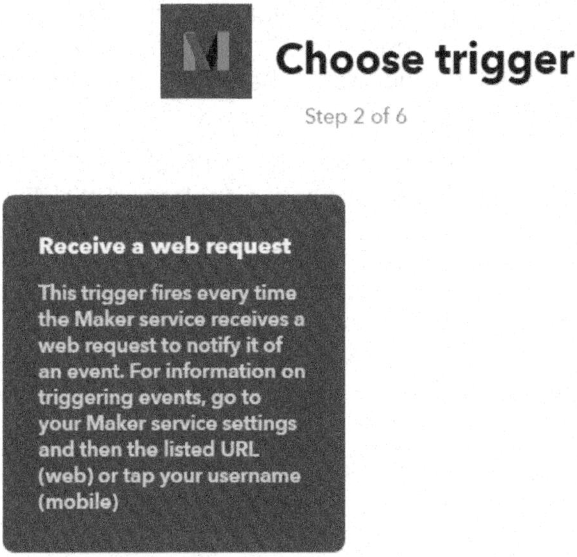

Choose trigger

Step 2 of 6

> **Receive a web request**
>
> This trigger fires every time the Maker service receives a web request to notify it of an event. For information on triggering events, go to your Maker service settings and then the listed URL (web) or tap your username (mobile)

Figure 9-6. *Choosing the trigger for the Maker service*

Click on this trigger, and you will be asked to give it a name. I am going to call mine button_pressed. Now click the Create Trigger button, and you will be asked to choose an action to perform by clicking on the blue "that" on the page (see Figure 9-7).

Want to build even richer Applets?

Figure 9-7. *Choosing the action on IFTTT*

This time, you can choose a service to perform an action when the Raspberry Pi triggers the web request. I decided to send myself an email. There is only one trigger for this service, so click on it and fill out the details of the email you will send to yourself (see Figure 9-8).

After you've set up things the way you want, click the "Create action" button. Then you can review your applet and click Finish.

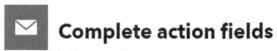

Complete action fields

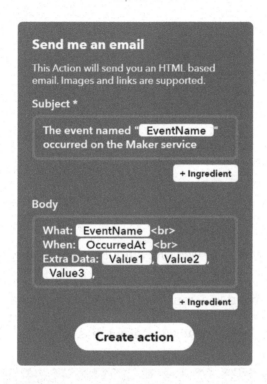

Figure 9-8. *The options for the email service*

To use this new applet from the command line, first enter the URL that you looked up before (from the settings of your Maker service) in a browser. You will see your key listed on the top. This page will also show you how to use the applet from a browser, and even gives you an example using `curl` (see Figure 9-9).

Your key is: ████████████████████

◄ Back to Channel

To trigger an Event

Make a POST or GET web request to:

```
https://maker.ifttt.com/trigger/ {event} /with/key/████ ████████
```

With an optional JSON body of:

```
{ "value1" : "       ", "value2" : "       ", "value3" : "       " }
```

The data is completely optional, and you can also pass value1, value2, and value3 as query parameters or form variables. This content will be passed on to the Action in your Recipe.

You can also try it with curl from a command line.

```
curl -X POST https://maker.ifttt.com/trigger/{event}/with/key/████ ██████████
```

Test It

Figure 9-9. *Maker service URL output with instructions for use*

So, for a one-off situation, you can just run this `curl` command on the command line, replacing *{event}* with the name of your event. If everything is configured correctly, you should get an email letting you know your event has been triggered (see Figure 9-10).

```
pi@raspberrypi:~/rclone-v1.34-linux-arm $ curl -X POST https://maker.ifttt.com/t
rigger/button_pressed/with/key/████ ████ ████████ ████
Congratulations! You've fired the button_pressed eventpi@raspberrypi:~/rclone-v1
.34-linux-arm $
```

Figure 9-10. *Triggering IFTTT with curl*

While this is fine for outbound triggers from your Raspberry Pi, it doesn't handle inbound requests, so you can trigger things locally from external requests. For that, let's set up a simple web server that can both send a request when a button is pressed and receive a trigger to turn on an LED. You could do this with more complex web servers with lots of features, such as Apache or even Lighttpd, but that's probably overkill for what most people need here. Instead, we will be using Flask, which is a Python web framework library for processing web requests. Because it's all written in Python, there are very few requirements and we can do everything we need to in a single script.

Flask should be already installed if you're running a recent release of Raspbian. If not, you can install Flask by first installing **pip**, the Python package management tool, like this:

```
sudo apt-get install python-pip
```

Then you can use **pip** to install the Flask library like this:

```
sudo pip install flask
```

We will also be using the requests library (which should be installed), and the pigpio library (which I showed you how to install in Chapter 7). To start, save the following Python script to a file on your Raspberry Pi. This script will start a web server and listen for a request to call the */light_switch* page. When that happens, it will turn an LED on or off. At the same time, it will wait for a button to be pressed and then send an outbound request to IFTTT to trigger the email event we set up earlier:

```
#!/usr/bin/python

## Import libraries
from flask import Flask
import requests
import pigpio
import time

## Define variables
app = Flask(__name__)
event = "button_pressed" #Event name from IFTTT
key = "your key here" # Key from IFTTT Maker service
# GPIO pins to use for the LED and button
led = 18
button = 24
pi = pigpio.pi()
# Setup the LED initially off
pi.set_mode(led, pigpio.OUTPUT)
pi.write(led, 0)
# Setup the button
pi.set_mode(button, pigpio.INPUT)
pi.set_pull_up_down(button, pigpio.PUD_UP)
# Debounce the button
pi.set_glitch_filter(button, 100000)
```

```
# This function will be called when the button is pressed
def button_callback(gpio, level, tick):
    url = "https://maker.ifttt.com/trigger/%s/with/key/%s"
    r = requests.post(url % (event,key))
    print str(r.status_code) + ":" + r.text
    print "Event %s triggered." % event

# This detects when the button is pressed and
# calls button_callback()
b_detect = pi.callback(button, pigpio.FALLING_EDGE, button_call
back)

# Set URL used to trigger hello()
@app.route("/")
def hello():
    return "Hello World! Waiting for input."

# Set URL used to trigger light_switch()
@app.route("/light_switch")
def light_switch():
    if pi.read(led) == 0:
        pi.write(led, 1)
    else:
        pi.write(led, 0)
    return "Light was been switched."

# Start the web server
if __name__ == "__main__":
    app.run(host='0.0.0.0', port=80)
```

Be sure to change the *event* variable to match the event name of your email applet from IFTTT. Also, change the *key* variable to match your IFTTT key that you found before. Don't forget to give the file executable permissions (chmod 755 *filename*) so you can run it as a script. Now connect the positive pin of an LED to GPIO pin 18 and the negative pin to ground. Connect a momentary switch between GPIO pin 24 and ground (see Figure 9-11).

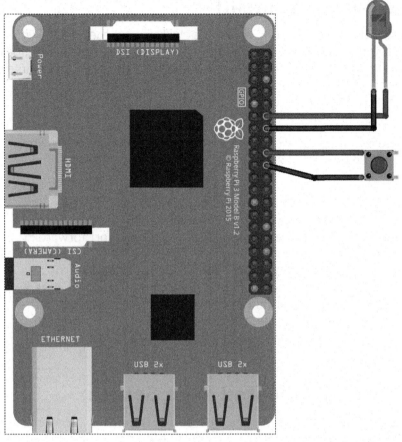

Figure 9-11. *Connecting an LED and switch to the Raspberry Pi*

Now run the script as sudo. From another computer on your network, you should be able to enter the IP address of your Raspberry Pi and get a basic web page that says "waiting for input." If you browse to the */light_switch* page (i.e., *http://xxx.xxx.xxx.xxx/light_switch*) your LED should turn on or off. If you press the button, it should send you an email.

Try It for Yourself

Try setting up an IFTTT applet that will request the */light_switch* page on your Raspberry Pi and turn the LED on or off. Perhaps you can use the Twitter service to request the page every time someone mentions you in a tweet. You could also use the Facebook service to request the page every time you're tagged in a photo.

Use whatever service you want for the "this" part of the applet and use the Maker service for the "that" part.

 Accessing Raspberry Pi from IFTTT
Make sure IFTTT can get to your Raspberry Pi from the internet. You might need to use port forwarding on your router to direct incoming traffic on port 80 to your Raspberry Pi's IP address. You might also need to use a different port number if you're already running another web server. You can search the internet for instructions on setting up port forwarding on your router model.

Run a Dedicated Web Server

Using Flask to run a simple web server is certainly convenient, but sometimes you need a more full-featured solution. Small dedicated web servers allow for better integration with other software, and better reliability and security than a standalone Python script. Also, many times, a web server can make it easier to integrate with cloud services like IFTTT, as we just demonstrated. Larger web servers like Apache can be run on the Raspberry Pi, but require more resources and may slow things down. Apache can also be quite complicated to configure and manage. Instead, I recommend using Lighttpd, which is easy to install and configure for most projects that will run on the Raspberry Pi.

Installation

You can install Lighttpd using `apt-get`:

```
sudo apt-get install lighttpd
```

Configuration for Python

If you want to run Python Common Gateway Interface (CGI) scripts on your web server, you will need to make a few changes. First, open the configuration file in nano:

```
sudo nano /etc/lighttpd/lighttpd.conf
```

Now add the following lines to the end of the file:

```
$HTTP["url"] =~ "^/" {
    cgi.assign = (".py" => "/usr/bin/python")
}
```

Save and close the file by pressing Ctrl-X, then Y, then Enter. Now you must enable the CGI module for Lighttpd by running the following command:

```
sudo lighttpd-enable-mod cgi
```

Then restart the Lighttpd server:

```
sudo service lighttpd restart
```

Test It Out

If you're running in a desktop environment, you can open a browser and point it to *http://localhost*. However, I find it's always best to test a web server from another host on the network to make sure everything is working. On another computer, use your Raspberry Pi's IP address instead of *localhost* (see Figure 9-12).

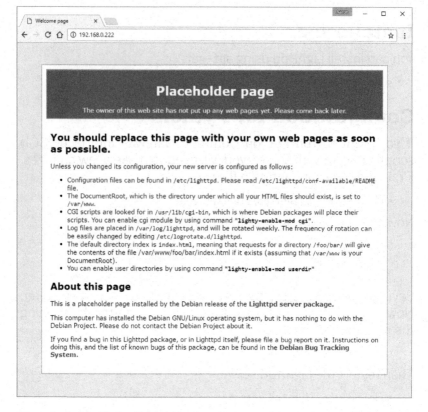

Figure 9-12. *Default Lighttpd web page*

You can now start developing your own web pages or use Python scripts to bring your projects to life across the internet. You can edit the *lighttpd.conf* configuration file to accept a Python file as the default page (i.e., *index.py*), or you can edit the *index.html* file and point to the location of the Python script you want to run, like this:

```
<html>
<head>
    <meta http-equiv="refresh"
    content="0; url=/cgi-bin/index.py" />
</head>
</html>
```

Just change the url= part to point to the relative location of your script. For more information on how to program Python scripts to work with web servers, take a look at the CGI module (*https://docs.python.org/2/library/cgi.html*). Keep in mind that */var/www/html* is the default directory that the web server looks in for files. So in the preceding example, the location of *index.py* is */var/www/html/cgi-bin/index.py*. Also, the web server runs as the user "www-data." Thus it's a good idea to assign ownership of any new files you want to use to that user. To do that, you can use what you learned in Chapter 6.

Roll Your Own

Instead of accessing cloud services on the internet, you can set up your own to run on your Raspberry Pi. This means you can store and share files inside your own network without using any external bandwidth, keeping your data private. There are many cloud storage software options that will run on the Raspberry Pi. OwnCloud and NextCloud are popular options, but are fairly difficult for new users to install. Instead, I will tell you about two other cloud storage services on your Raspberry Pi that are much easier to install and require fewer resources to run.

Nimbus

Nimbus (*https://cloudnimbus.org/*) is simple cloud storage software designed to run on the Raspberry Pi. As of this writing, it is still in beta, but it is quite capable and ready for home use. It also has a client for Windows that will sync your files for you. First you need to create a directory for the Nimbus files:

```
mkdir /home/pi/nimbus
cd /home/pi/nimbus
```

Now you can download the software to your system by using wget:

```
wget http://cloudnimbus.org/dist/0.6.2-BETA/nimbus-0.6.2-
BETA.tar.gz
```

This file will have to be extracted before you can use it. You can do that with the tar command:

```
tar -zxvf nimbus-0.6.2-BETA.tar.gz
```

Nimbus comes with a script to install all the necessary software that may not already be installed on your Raspberry Pi. Run that script from the command line as **sudo**:

```
sudo ./install_helper_programs.sh
```

Now start Nimbus by running the *nimbus.sh* script:

```
./nimbus.sh start
```

By default, Nimbus uses port 8080, so if you're running in the desktop environment, you can open a browser and point it to *http://localhost:8080* to access Nimbus. If you're not running the desktop, from another computer you can browse to the IP address of your Raspberry Pi and add *:8080* at the end (see Figure 9-13).

Figure 9-13. *The Nimbus first-run welcome page*

The first time you access your Nimbus home page, it will ask you to create an account. Once you do this, you can log in to the service and begin loading and sharing files (see Figure 9-14).

If you're happy with the way things are working, and you want to run Nimbus every time your Raspberry Pi boots up, you can run the included script to set up Nimbus as a service on your system:

```
sudo ./add_to_startup_programs.sh
```

You can also add an external USB hard drive to expand the amount of storage available. For details on how to do this, check out the Nimbus website (*https://cloudnimbus.org/*).

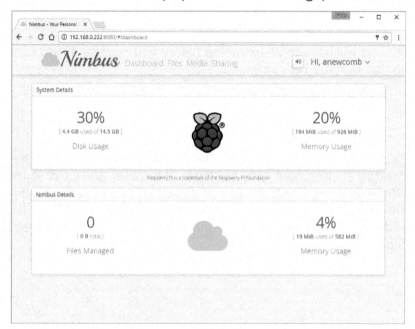

Figure 9-14. *The Nimbus home screen*

Tonido

Tonido is another cloud storage service that you can run on your Raspberry Pi. It was designed to run on SBCs and has been around for quite a while, so the software is mature. With Tonido, you can store, share, and stream files, and there are clients available for PC, Android, and iOS. Even though the public Tonido site does not store your password or any files, it does relay you to your private server when you're not connected to your network. Tonido is a single-user system, but you can create guest accounts to give other people access to specific files. The installation is very similar to the way Nimbus is installed. First you need to create a directory for the Tonido files:

```
mkdir /home/pi/tonido
cd /home/pi/tonido
```

Now you can download the software to your system by using wget:

```
wget http://patch.codelathe.com/tonido/live/installer/armv6l-
rpi/tonido.tar.gz
```

This file will have to be extracted before you can use it. You can do that with the **tar** command:

```
tar -zxvf tonido.tar.gz
```

Now start Tonido by running the *nimbus.sh* script:

```
./tonido.sh start
```

By default, Nimbus uses port 10001, so if you're running in the desktop environment, you can open a browser and point it to *http://localhost:10001* to access Tonido. If you're not running the desktop, from another computer you can browse to the IP address of your Raspberry Pi and add *:10001* at the end (see Figure 9-15).

Figure 9-15. *The Tonido first-run welcome page*

Follow the prompts to create your account. This will also create a public web address that you can use to make it easier to access your Tonido file-sharing server. Once you're logged in, you can start uploading and sharing files (see Figure 9-16).

Figure 9-16. *Tonido user home page*

Notice that some of the built-in applications that allow things like searching and connecting from mobile applications may be in a suspended state at first. To fix this, click on the warning that appears on the user's home page. You can see it highlighted with the blue box in Figure 9-16. Then click individually on each application and select Resume. In my case, clicking the Resume All Applications button did nothing.

You can start Tonido automatically when your Raspberry Pi boots up by adding the start command to your *rc.local* file, as shown in Chapter 6.

Why This Matters for Makers

Knowing how to set up and use cloud services can open up a whole new world of possibilities for a Maker. Not only does it allow you to transfer and store files, but it also allows you to add internet control of your project. As the popularity of cloud-connected systems grows, you will be able to build your own IoT devices using the Raspberry Pi.

10/Virtual Raspberry Pi

Makers are busy people and don't always have a Raspberry Pi sitting in front of them to experiment with, or perhaps they aren't sure if they will be able to get over the "Linux hump" and become proficient enough to justify buying a Raspberry Pi or other SBC. For these cases, I created a Virtual Raspberry Pi that runs in Oracle VirtualBox, which you can use to explore Linux and test things out. In fact, many of the images and exercises in this book were made using my Virtual Raspberry Pi. This chapter explains how to run a full Raspberry Pi environment on top of Windows, macOS, or Linux for those times when you don't have the physical system on hand.

Before you get started, there are a few things to understand about the virtual environment. First, since this is not a physical system, there are no physical GPIO pins to interact with. There is no way to hook up any real-world components to the virtual environment. So if you thought you might use this to test code that interfaces with I²C or SPI devices, be aware that there really is no way to do that. Second, the current emulation environment is limited to 256 MB of memory, so the system will not be very speedy. Third, the Virtual Raspberry Pi image is actually virtualized twice. There is a main guest image running stock Debian Linux, and then I use QEMU to virtualize an ARM environment for the Raspbian disk image inside of Debian. This doesn't seem to have any ill effects except that it made routing access to the internet a little more cumbersome. Luckily, I have taken care of all of this for you.

Requirements

- A computer running a recent version of Windows, macOS, or Linux
- 15 GB of free disk space

Installation

Download and install Oracle VirtualBox from virtualbox.org.

Download the Virtual Raspberry Pi image file from Google Drive (*http://bit.ly/2o98Dg1*) and save it somewhere you will remember. This is a 5 GB file, so it could take quite a while to download.

Open VirtualBox and click on File→Import Appliance (see Figure 10-1).

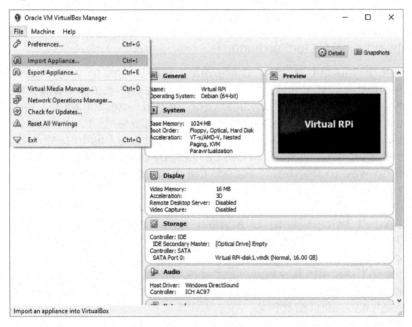

Figure 10-1. *Importing a preconfigured VirtualBox image*

In the file search window, find and select the image file you downloaded from Google Drive. This will import and configure the Virtual RPi system inside of VirtualBox for you.

Usage

To launch the Virtual RPi machine, select it from the list on the left and click the Start button. VirtualBox will then run the system in a separate window. You will notice that the Linux bootup messages appear twice in a row. This is due to the fact that the Raspberry Pi is being virtualized inside of a virtual Debian installation, as I mentioned earlier. Once everything is finished booting, you will be at the Raspberry Pi command prompt (see Figure 10-2).

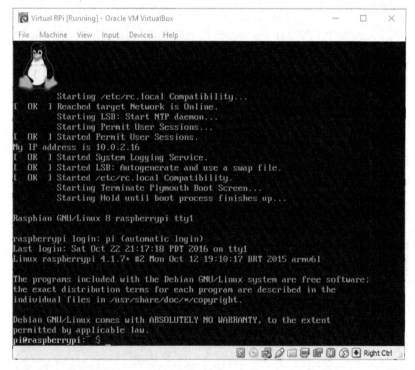

Figure 10-2. *The console of the Raspberry Pi image after it has booted up*

The desktop does not run by default, but you can start it by typing **startx** on the command line.

To shut down the Virtual RPi machine, you must first shut down the Virtual Raspberry Pi by rebooting it. Then shut down the

Debian environment. To shut down the Virtual Raspberry Pi, type:

```
sudo shutdown -r now
```

The Virtual Raspberry Pi will not reboot. Instead, you will be left on the Debian desktop. To shut down the Debian environment, simply click on the menu icon in the bottom-left corner, choose Logout, and then select Shutdown (see Figure 10-3).

 Don't Shut Down

Using `shutdown -h` is not recommended on this virtual system because it will not return you to the Debian desktop. If this happens, you can press Ctrl-Alt-F to exit full-screen mode and close the emulator window.

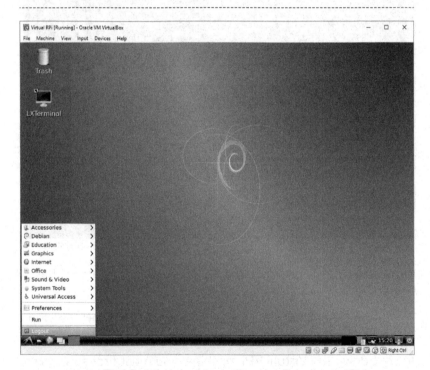

Figure 10-3. *The Debian desktop after the Raspberry Pi image has been rebooted*

If you're stuck in the Raspberry Pi image and would like to get access to the Debian desktop, you can press Ctrl-Alt-F to exit full-screen mode. To toggle the mouse capture mode, press Ctrl-Alt when you're not in full-screen mode.

Why This Matters for Makers

Because you can! Sometimes it's fun just to try things out to see how they work. While the lack of memory and physical GPIO pins might not make this the preferred way to learn about the Raspberry Pi, it can still be fun to poke around and see how things work on the system. You can use this virtual installation of the Raspbian distribution as a reference to test the syntax of a program, verify where software resources are located, and get material for documenting an instructional blog post when you don't have a physical system in front of you. You can also learn a little about virtual computing in the process, or even use this virtual Raspberry Pi to write a book!

A/Linux Background

This appendix will cover the history of the Linux operating system, as well as some general questions that often come up when Makers start considering using Linux in their projects. Linux is used all over the world on many different platforms. For businesses, it's the most preferred operating system for running complex transactions in their data centers. It is also used as the base operating system in all Android phones, making it the most used operating system in the world today. In addition, Linux is used in embedded devices like network routers, point-of-sale terminals, medical devices, set-top boxes, and digital TVs. Of course, Makers use it all the time as the operating system for SBCs like Raspberry Pi, BeagleBone Black, C.H.I.P., Pine 64, and Onion platforms. And because there are so many versions available, you can easily bring old hardware back to life by installing Linux on it.

Brief History of the Original Maker Operating System

I think it's fair to say that Linux started as (and still is) the ultimate Maker project. Before Linux, the easiest and most accepted way to run a Unix-type operating system was to pay for one of the commercial products available in the marketplace. Unix was created by Ken Thompson and Dennis Ritchie while they were working at Bell Labs (now known as AT&T) and was first released in 1970. At that time, operating systems were designed to run on specific hardware platforms. General-purpose computing was still in its infancy; you couldn't take the software from one hardware platform and run it on a different one. As Unix became more popular, similar operating systems were created, including Berkeley Software Distribution (BSD) and SunOS.

For the next 20 years, the way companies acquired and ran Unix in their data centers was to buy a license from AT&T, HP, IBM, Sun Microsystems, or another vendor that sold a version of Unix. For the most part, these companies developed their operating systems in a vacuum. As you can imagine, they wanted to protect their investments and research from getting into the hands of their competitors. Even though they made improvements and fixes to the code over time, the end user was at the whim of the vendor in terms of which fixes got implemented first and which improvements bubbled up to the top of the list of features for the next release of the software.

Where did this leave the Maker? Pretty much out in the cold. Even though the Unix operating system was very powerful and became more versatile over time, the software licenses were too expensive for most individuals to afford. The computer hardware needed to run Unix was also very expensive, so even if an individual did want to run Unix for a project, they had to rent time on a server from a company that had these systems installed. Imagine if a Raspberry Pi that costs $35 today instead cost $35,000, and Raspian was only available if you first paid $10,000 just for the privilege of running it. Or imagine if the Raspberry Pi Foundation wanted to charge you $50/hour just to use a Raspberry Pi for a little while.

The concept of sharing access to computers and renting that time out to others became known as *time sharing*. It was a more efficient use of an expensive piece of equipment and allowed companies to recoup some of their investment. However, if you were lucky enough to be a student at a university that had access to these time-sharing systems, you could use them for free. This got a couple of people thinking.

Figure A-1. *Time sharing ad from 1970 (image from HP Computer Museum (http://www.hpmuseum.net/))*

When Richard Stallman started the GNU's Not Unix (GNU) Project in 1983, he was working for the Massachusetts Institute of Technology (MIT) as a research assistant in their Artificial Intelligence Laboratory. He was frustrated by the restricted computer access imposed by the lab, as well as the increasing trend toward developing proprietary software that couldn't be modified or distributed to others. In one case, he had modified the software for a laser printer to be able to notify the owner when the print job was complete. However, when a new printer was installed with proprietary software, he was restricted from making the same changes, which meant people had to walk up and down stairs multiple times to check on their print jobs or wait by the printer until they were done.

The GNU Project's goal was (and is) to allow people access to software that could be freely used, shared, and distributed in a collaborative way. That meant that if you wanted to improve the software in some way or fix something that wasn't working, you could just do it without having to ask permission or hack the code. You could also pass on your changes to others so that they could learn from them and use them in their own software. The GNU Project made their own versions of utilities commonly found on Unix systems that were necessary for the development of software, like a text editor (Emacs), code compiler (GCC), and code debugger (GNU debugger), as well as common tools like ls, grep, and make. In many respects, Stallman and the others who worked on the GNU Project were Makers in their own rights. Just like modern-day Makers, they believed people should hack, make, and share their projects with other people who have similar interests.

In an effort to codify these beliefs and make them applicable to the software he and others were developing, Stallman came up with the GNU Public License (GPL). You may have seen this when you installed some software on your computer or phone. Large parts of the Android operating system, WordPress, GIMP, VLC Media Player, and even the Linux kernel itself (we'll get to that in a minute) use the GPL as their software license. The GPL states that you may modify, copy, and redistribute the software, but if you do, you must keep the GPL license in the new or copied software. This ensures that all users of the software get the same rights no matter how many times the software changes.

This license and several other software licenses created around the same time led to what we now know as *open source software*. While proprietary, or *closed source*, software usually imposes restrictions on how the software can be used to protect the company or author, open source software aims to protect the author while providing rights and protections to the user as well. Even though there are many different open source licenses in existence today, the GPL is still the most popular. And this belief that software should be free to be modified and improved eventually spilled over into hardware development as well. Some of the most popular platforms Makers use for building projects like Arduino—including the RepRap project, Lulzbot, Beagle-Bone Black, and most of the Raspberry Pi—are considered *open source hardware*. This means their specifications are published publicly and people are free to modify their designs.

Try It for Yourself

Although not strictly required to be offered at no charge, many free software programs use an open source license. Try opening some of your favorite programs and look in the About section under the Help menu and see what license your software uses. You can also search the internet for this information. You might be surprised how much software uses or is based on open source licenses. It represents quite a change from the early days of computing.

Linus Torvalds

Even though work was progressing nicely on the GNU Project's programs, tools, and utilities, they were still lacking a decent Unix kernel. In 1991, a computer science student at the University of Helsinki named Linus Torvalds had just ordered a new computer and a copy of another Unix clone called Minix. The source code for Minix was available, but it was not allowed to be modified and redistributed. Like all good Makers, Torvalds took this as a problem to be solved. He believed there should be a freely available Unix-like operating system that ran on the still-new x86 computer platform. Here is his note to the Minix community announcing his new operating system.

> Hello everybody out there using minix -
>
> I'm doing a (free) operating system (just a hobby, won't be big and professional like gnu) for 386(486) AT clones. This has been brewing since april, and is starting to get ready. I'd like any feedback on things people like/dislike in minix, as my OS resembles it somewhat (same physical layout of the file-system (due to practical reasons) among other things).
>
> I've currently ported bash(1.08) and gcc(1.40), and things seem to work. This implies that I'll get something practical within a few months, and I'd like to know what features most people would want. Any suggestions are welcome, but I won't promise I'll implement them :-)
>
> Linus (torvalds@kruuna.helsinki.fi)

PS. Yes - it's free of any minix code, and it has a multi-threaded fs. It is NOT portable (uses 386 task switching etc), and it probably never will support anything other than AT-harddisks, as that's all I have :-(.

— Linus Torvalds[1]

This new operating system would later become known as Linux, which is pronounced (['liːnəks]). The name is a combination of "Linus" and "Unix." It is interesting to note that, at the time, it was just a fun project to work on. Torvalds didn't think it would be "big and professional," or portable, or support a wide variety of peripherals. My, how things have changed. Today, Red Hat and others have built enormous companies around offering professional software to businesses based on Linux. It is also one of the most portable operating systems in existence today, running on almost every computer architecture available. Not to mention, Linux supports tens of thousands of devices, both new and old. Many times, they are supported with open source drivers that are already in the system so you don't have to install software.

Linux became popular quickly because it didn't cost anything to run, was easy to obtain, and was easily modified and improved. You could make a comparison to the Arduino ecosystem that exists in the Maker community today. Arduinos are very inexpensive and easy to order online, and if you don't like the way your Arduino Uno is laid out, you can build your own from scratch to suit your needs. As Linux grew in popularity, it also developed a multitude of communities thanks to the "open" nature of the development model. Torvalds not only received requests for changes to the code, but he also got actual code samples from programmers who were trying out the operating system. Since the code was openly available for people to read and study, the development cycle for new features and bug fixes was much quicker than with the previous proprietary Unix systems. With the amount of changes being submitted, it quickly became necessary to organize things so that changes could be reviewed before they were included in new versions of Linux.

1 Linus Torvalds. "What Would You Like to See Most in Minix?" Usenet group *comp.os.minix*, August 25, 1991.

Linux is still maintained and improved in much the same way today. Torvalds and a handful of others called "Linux kernel maintainers" still govern this process, and it is still open to anyone who wants to improve the code or add new functionality. Don't have a driver for that new WiFi module? You can write one yourself. Discover a bug in the way the operating system boots up? You can submit a patch to fix it. It truly is the original Maker operating system.

The Linux Kernel

I've mentioned the Linux kernel a few times now, and as you learn about using Linux for your projects, you will see it referenced online in tutorials and forums. Similar to a seed at the center of a nut or fruit, the kernel is the core program that manages the functions of the operating system. It is usually the first program that is run when the system starts up. It sits between the applications and the hardware components and governs how and when those components can be accessed. Without the kernel, applications wouldn't be able to run because they wouldn't know how to access the CPU, memory, storage, and other hardware that makes up the computer. The kernel also acts like a traffic cop, preventing applications from "running into" each other as they request the same resources. Modern operating systems have thousands of programs running simultaneously. If you didn't have a kernel, you could really only run one program at a time without causing problems. As you can tell, the Linux kernel is the most important part of the operating system. Because it's so important, it is loaded into a secure part of system memory so that it's protected from tampering and changes.

There are many versions of the Linux kernel. It's updated frequently by the kernel maintainers and can be customized to include all or just some parts depending on the needs of the system. If your system doesn't have much memory, for example, a developer could take out all the parts they didn't need in an effort to make the kernel smaller. If your system isn't very fast, a developer might want to take out the parts that don't need to run all the time in order to make the kernel more efficient. Some individual users even go so far as to make custom versions of the kernel by changing the included components or modifying certain parameters in an effort to make the system work the way they need it to. This process of compiling your own Linux kernel is mostly unnecessary for the Maker, as this work is already done by those who make and distribute the devices we use, like the Raspberry Pi.

Because the Linux kernel is so configurable, it often includes all the software you need to make your components visible to the operating system. This software is also called a *driver* and, depending on the manufacturer, might be proprietary or open source. If it's open source, it can be included in the kernel and that makes adding components to your Linux system a lot easier. For example, when you connect almost any generic mouse or keyboard to your Linux-based system, it will be automatically detected and configured. With other operating systems, like Microsoft Windows, the driver software has to be installed from the internet or from local storage before the device can be used. Sometimes the drivers for older devices can be hard to find or become unsupported. This is less likely to happen with an open source driver in the Linux kernel because once it's written, it can always be referenced again whenever it's needed. This is one reason open source software should be important to Makers, and I will talk more about this a little later.

Distributions

I mentioned in the introduction that I would focus mainly on the Raspian distribution of Linux. But what is a distribution anyway? At face value, the word *distribution* makes it sound like someone is sharing their stuff with a lot of other people. That is not very far off the mark. Anyone can customize an operating system based on Linux by adding and changing the various programs that make it useful. It can then be packaged up in an easy-to-install format that ensures that every installation will be exactly the same. This collection of preconfigured software is called a distribution. A distribution of Linux can be critical for deployment when you are installing more than one system at a time. For example, let's say you wanted to install Linux on 100 servers at work. Without using a distribution you would have to compile your kernel, choose what desktop you wanted to use, install all your software, and configure all your services for each server one by one. By using a distribution, you can greatly simplify this process by installing a preconfigured set of software on each server, ensuring that each server's installation will be exactly the same as the one before.

Distributions also allow individuals and companies to publish their particular flavor of Linux to the world. Once all the software is in place, an installer program is added to make installation easy and less time-consuming. Then, all the files that make up the operating system are packaged up into a single file to make it easier to download. Like the Linux kernel itself, distributions are maintained by a group of people and updated on a regular basis to make sure the software stays current, fix bugs, and add new features.

This idea of creating your own distribution of an operating system is fairly unique to Linux. In order to get a customized version of other operating systems like Windows or macOS, you would need to convince Microsoft or Apple to make them for you. Because Linux is based on open source software, you can just make your own choices about what software you need for the task you're trying to accomplish. Just as you would choose a set of tools in your workshop for a given task, you can choose a Linux distribution that best meets the needs of your project.

Examples of popular Linux distributions include Linux Mint, Debian, Ubuntu, OpenSUSE, and Arch. Some companies, like Red Hat, release a commercial distribution and a community distribution (Fedora). Some distributions are derived from another distribution. For example, Ubuntu and Raspbian are derived from Debian, whereas Linux Mint is derived from Ubuntu. This means that they start with one distribution as a base and then make changes to the software or look-and-feel and redistribute as a standalone system. There are also many distributions that are specialized for a given task or system type. Ubuntu Studio is built to appeal to audio, video, and graphical designers. Tiny Core Linux is a full desktop operating system that occupies as little as 16 MB, runs entirely in memory, and loads from a USB thumb drive or CD. GParted Live is another distribution that runs from external storage. It helps diagnose problems and make changes to storage hardware.

There are many distribution of Linux that run on the Raspberry Pi as well. Raspbian is specifically designed to run on the Raspberry Pi and is the most popular choice of users today. It's also officially supported by the Raspberry Pi Foundation. Other Raspberry Pi distributions include:

Ubuntu Mate (https://ubuntu-mate.org/raspberry-pi/)
> A version of Ubuntu Mate optimized to run on Raspberry Pi 2 and Raspberry Pi 3. It's great if you want to use Raspberry Pi as a desktop or are already familiar with Ubuntu.

OpenElec (http://libreelec.tv/)
> An embedded operating system built around Kodi, the open source entertainment media hub. This is for users who want to use the Raspberry Pi as a media center only.

Open Source Media Center (https://osmc.tv/download/)
> Another media center software distribution that is based solely on open source software.

PiNet (http://pinet.org.uk/)
> A distribution that runs in a network topology to make it easier for educators to use Raspberry Pi in the classroom.

Try It for Yourself

There are thousands of distributions of Linux available for download. You can see some of the most popular ones by taking a look at the DistroWatch website (*http://distrowatch.com*). They track the popularity of Linux distributions as well as the latest updates both for the distributions and popular open source software packages. Click on some of the distributions listed to see what they do and how they are different.

How Open Source Software Works

I've talked a little about open source software and how it was critical to the development of Linux, but it is important for Makers to understand how open source software actually works in order to appreciate how it could benefit a project and how to avoid potential problems.

Open source software is different than proprietary or closed source software in many ways. First and foremost, as the word *open* implies, the source code for the software is available for anyone to look at and inspect. For example, you might be wondering why a program does x when you do y, but not when you do z. Perhaps you want to know because the software is working incorrectly, or perhaps you are a programmer and would like to implement a similar algorithm, function, or programming technique in your software. With proprietary software there is no way to know for sure. You would need to ask the company that wrote the software to file a bug report that may or may not get addressed. With open source software, however, if you know a bit of programming, you can actually look at the code and see what is going on. You may or may not be able to understand all the code or know how to fix the issue, but at least you aren't operating behind a wall of secrecy and uncertainty.

Because the code is publicly available, some people wrongly assume that this is an enormous security risk. They think that it would be easy for bad actors to take the code and do something malicious with it like insert a virus or malware that could harm people's computers and devices. While it's true that the "openness" of open source software might make it an inviting target, it's also the very thing that keeps this from becoming a real problem. Because the code is open, it is constantly under inspection. The release and acceptance of new code can be reviewed and governed by a team of experts from around the world, and more people can run tests and nightly builds to see how the software is working. Bugs and security risks can be identified and corrected much more quickly than with proprietary software.

Second, because the code is available, this naturally invites contributions from the community. If there's a problem with the software, you can correct it yourself by submitting a bug report or patch that fixes the issue. If programming is not your thing, there are other great ways to get involved as well. Open source projects are always looking for people to help with non-programming tasks like documentation, translation, marketing, website development, and community relations. Getting involved in an open source project is a great way for people who are thinking of a career in software development, marketing, or developer relations to get some experience.

Third, open source software can be shared and distributed to others that need it without breaking the law or violating some sort of license agreement. In fact, this behavior is generally encouraged. This is in stark contrast to what you may typically think of when it comes to sharing content. Media organizations like the RIAA or MPAA spend vast amounts of time and money discouraging people from sharing their constituents' legally protected content, but open source projects don't have those restrictions and post their content on sites that encourage collaboration and sharing, like GitHub and SourceForge. File sharing software like BitTorrent can be used to make downloading easier because there are no legal issues to contend with.

Open source software really benefits the Maker community because ideas and projects can be implemented, shared, and improved more easily when software is readily modified and readily available. However, there are a few things to look out for when using open source software in your project. Just like all software, open source software can exist in various stages of development. An open source project might run out of steam and not be under active development. The risk here is that new features and functionality that you need for your project might never be implemented. Open source software is also more likely to be released early in the development cycle so it can be tested and improved. The risk here is that there may be more bugs in the code and that the software might change in functionality more quickly than with stable code. Imagine you're using some brand new software that is in the early stages of development and it solves a problem in one of your projects. You get everything working just the way you want it, but the next time you update your software, things go haywire. As it turns out, the code you were using was modified in a way that broke the way you were using it and you need to spend time figuring out how to change things to make it work again. Luckily, with open source software, these types of problems are easy to fix, but it is good to be aware of the maturity of the software you are using to minimize the impact to your project.

Fourth, it is good to be aware of what kind of open source license your software uses. I have already mentioned the widely used GPL, but there are hundreds of other open source licenses as well. Some of them are very permissive and others come with many restrictions. Most of them allow for commercial use, private use, modification, and distribution. They also almost always protect the developer from liability if the software is misused or fails to work. Some of them (like the GPL) require that changes to the code be made available as open source software, and some (like the BSD) do not. If you have ever used macOS on an Apple product, you may have noticed that when you use the command line for anything it feels like Linux. That is because macOS is based in part on a UNIX variant called FreeBSD, which (as you might expect) uses the BSD license. Because the BSD license doesn't require changes to be made available as open source, Apple is free to take the code and use it, change it, and improve it without sharing back to the community. While they are certainly within their right to do this, wouldn't it be nice to have those changes benefit the community that started them in the first place? There may come a time when you might want to fork a project or make changes to some code that is under an open source license. It's important to be aware that you may be required to keep your code under the same license or contribute the code back to the community. You can find out more about the different open source licenses at opensource.org.

Single-Board Computers Versus Microcontrollers

Although it is not strictly related to the history of Linux, this is probably a good place to talk a little about the similarities and differences between single-board computers (SBCs) and microcontrollers. You hear both of these terms quite often in the Maker community, but it can be difficult to determine when and why to use one or the other.

Microcontrollers are chips called integrated circuits (ICs) that contain a processor, a small amount of memory, and some ability to connect to things via input and output connections. These chips are often built into a platform that breaks out the various connections into pins to make it easier to communicate with other devices and modules. Sometimes these platforms include a USB port and controller so that users can connect directly to a computer and upload the firmware that runs on the chips. Examples of microcontroller platforms include Arduino, Teensy, ESP8266, and the Ti MSP430 LaunchPad.

SBCs are, as the name suggests, complete computer architectures that just happen to be built into a single PCB board. They typically include discrete components on the board like memory, a storage controller, USB connections, video and audio capabilities, and networking. They also usually run a complete operating system like Linux. Examples of SBC platforms include Raspberry Pi, BeagleBone, Odroid, and C.H.I.P. See Figure A-2 for an example of a Raspberry Pi and an Arduino.

Figure A-2. *The Raspberry Pi 3 (left) and the Arduino Genuino Uno (right)*

Both SBCs and microcontrollers are very powerful platforms for building things because both can be used to connect to and control the physical world. In recent times, they have both become quite small in size, which makes them useful in small projects where you need to hide them away inside a project box or inside an existing enclosure. Another similarity is that they can be modular in that you can add functionality by plugging in a device or board. Raspberry Pi calls these *hats*, while Arduino calls them *shields*.

Microcontrollers are great when you need to do a simple specific task (or set of tasks) repeatedly and reliably. Compared to SBCs, microcontrollers don't have a lot of memory, so the number of instructions and program size they can process at one time are limited. However, microcontrollers usually don't have a lot of peripherals and other programs competing for resources, so this makes them very fast at doing what they do. Since the firmware for a microcontroller is stored in memory and always runs when they boot up, they will always behave the same way every time you turn them on. Microcontrollers are great at running strips of LED lights or continually polling sensors to gather data. They are also good at sending instructions repeatedly in the case of motors for a CNC machine or 3D printer.

SBCs can be used more like a regular computer. Because they have more memory and dedicated storage devices, and run a complete operating system, they can be used to do multiple tasks at once. However, since there is more going on at the same time, you might have to deal with resource contention. Also, since the operating system is stored on storage and not entirely in memory all the time, you need to take care not to power off the system abruptly or you could end up with a corrupted filesystem, leaving your SBC unable to boot up or severely unstable. Instead of having to upload the firmware every time you want to make changes to the system, you can use SBCs in a normal desktop environment or run them from a console that allows you to make changes to programs and features while the system is running. Another advantage of SBCs running Linux is that you can write programs in many different programming and scripting languages very easily. Although Python is widely used for building projects with the Raspberry Pi, you could just as easily use Perl, Java, Go, or C.

Why This Matters for Makers

Makers can benefit from powerful, versatile environments for their projects. The freedom to be able to choose from many different types of software and programing tools means that they can pick the one that is right for them. Makers also need access to software on the fly as their projects change and develop, and many projects are developed on a shoestring budget. Linux, a community Maker project in its own right, gives Makers the freedom to work on their projects using the latest technologies without additional cost and simultaneously provides a customizable set of software packages that are just right for their requirements. Many open source projects started because someone had an itch they wanted to scratch. In much the same way that a Maker's project can start out as an individual effort and grow into a big business, open source software often starts as a pet project and develops into something that thousands of people use on a daily basis, sometimes without even realizing it. Open source software is at the heart of Linux, and it makes sense that Makers would develop a preference for using these tools similar to how they use a screwdriver, hammer, or 3D printer. Using Linux as part of your project may seem like a daunting task at first, but with just a few hints and tips, I think you will come to enjoy the freedom and endless possibilities it provides.

Index

R

Raspberry Pi, 1
  booting, 13
  cloud services, accessing (see
    cloud services)
  expanding filesystem on, 14
  GPIO pins in (see GPIO pins)
  hostname for, changing, 127-131
  IP address for, determining,
    95-99
  localization options for, 15-18
  multimedia, playing (see multi-
    media)
  password for pi user, 19, 32
  performance, monitoring,
    138-145
  power supply for, 13
  SSH server for, 99-101
  virtual, 219-223
  VNC server for, 108-110
Raspbian disk image, 2, 232-234
  downloading, 2
  uncompressing, 3-6
  writing to SD card, 7-13
raspi-config tool, 14, 93-95, 100
rc.local file, 131-133
rclone utility, 196-199
RealVNC, 110-116
rebooting, 90-91
regular expressions (see grep command)
relative paths, 54
relay module example, 176-177
remote command-line access,
  99-107
remote desktops, 108-117
  client for, on Linux, 114-116
  client for, on macOS, 112-114
  client for, on Windows, 110-112
  server for, on Raspberry Pi,
    108-110
resistors, when needed, 175
rm command, 60
root directory, 29, 29
/root directory, 30
root user, 32, 52

router, determining IP address
  from, 97-98
RPi.GPIO module, 170
/run directory, 30

S

SBC (single-board computer), 1,
  237-239
  (see also Raspberry Pi)
/sbin directory, 31
scheduling jobs, 165-168
SCP (Secure Copy), 117-126
scripts (see CGI scripts) (see shell
  scripts)
SD card
  downloading operating system
    to, 2-13
  requirements for, 1
Secure Copy (SCP), 117-126
Secure File Transfer Protocol
  (SFTP), 117-126
secure shell (see SSH client) (see
  SSH server)
security
  password for pi user, 19, 32
  root user, limiting use of, 33
  sudo command, 35
  WPA (WiFi Protected Access),
    77-79
semicolon (;)
  ending a command, 165
  running multiple commands,
    163
services, 36-38
SFTP (Secure File Transfer Protocol), 117-126
sh (Bourne shell), 26
sh command, 27
shell, 25-27
  (see also command line; termi-
    nal)
shell scripts, 26-27
  output from, capturing, 153-155
  playing media files from,
    193-194
  starting on bootup, 131-133

About the Author

Aaron Newcomb has been a Maker since he was old enough to hold a screwdriver and has been using Linux since 1997. He has worked in the IT industry for companies like New Relic, NetApp, Oracle, Sun Microsystems, and Hewlett Packard. He cohosts several shows about technology for TWiT LLC, including *FLOSS Weekly*, *All About Android*, *This Week in Google*, and *The New Screen Savers*. In 2012, he founded the nonprofit organization Benicia Makerspace, where he currently serves as president and executive director when he is not busy at work.

Colophon

The image on the cover of *Linux for Makers* is Tux, the Linux penguin, lovingly cradling the Raspberry Pi logo, symbolizing the synergy between the hardware and software sides of the Maker movement.

The cover image is by Brian Jepson. The cover and body font is Benton Sans; the heading font is Benton Sans; and the code font is TheSansMonoCd.

CPSIA information can be obtained
at www.ICGtesting.com
Printed in the USA
LVOW06s1113241017
553517LV00018B/9/P